Technology and Engagement

Technology and Engagement

Making Technology Work for First-Generation College Students

HEATHER T. ROWAN-KENYON

ANA M. MARTÍNEZ ALEMÁN

MANDY SAVITZ-ROMER

Rutgers University Press

New Brunswick, Camden, and Newark, New Jersey, and London

Library of Congress Cataloging-in-Publication Data

Names: Rowan-Kenyon, Heather T., 1973– author. | Martínez Alemán, Ana M., author. | Savitz-Romer, Mandy, author.
Title: Technology and engagement : making technology work for first generation college students / Heather T. Rowan-Kenyon, Ana M. Martínez Alemán, Mandy Savitz-Romer.
Description: New Brunswick : Rutgers University Press, 2018. | Includes bibliographical references and index.
Identifiers: LCCN 2017011904 (print) | LCCN 2017039578 (ebook) | ISBN 9780813594217 (E-pub) | ISBN 9780813594231 (Web PDF) | ISBN 9780813594200 (hardback) | ISBN 9780813594194 (paperback)
Subjects: LCSH: First-generation college students—United States. | College preparation programs—United States. | Educational technology—Study and teaching (Higher)—United States. | BISAC: EDUCATION / Higher. | EDUCATION / Computers & Technology. | COMPUTERS / Social Aspects / Human-Computer Interaction. | TECHNOLOGY & ENGINEERING / Social Aspects. | COMPUTERS / Educational Software.
Classification: LCC LC4069.6 (ebook) | LCC LC4069.6 .R68 2018 (print) | DDC 371.33—dc23
LC record available at https://lccn.loc.gov/2017011904

A British Cataloging-in-Publication record for this book is available from the British Library.

∞ The paper used in this publication meets the requirements of the American National Standard for Information Sciences—Permanence of Paper for Printed Library Materials, ANSI Z39.48–1992.

www.rutgersuniversitypress.org

Manufactured in the United States of America

We would like to dedicate this book to first-generation college students everywhere, and especially those at Millbrook University.

Contents

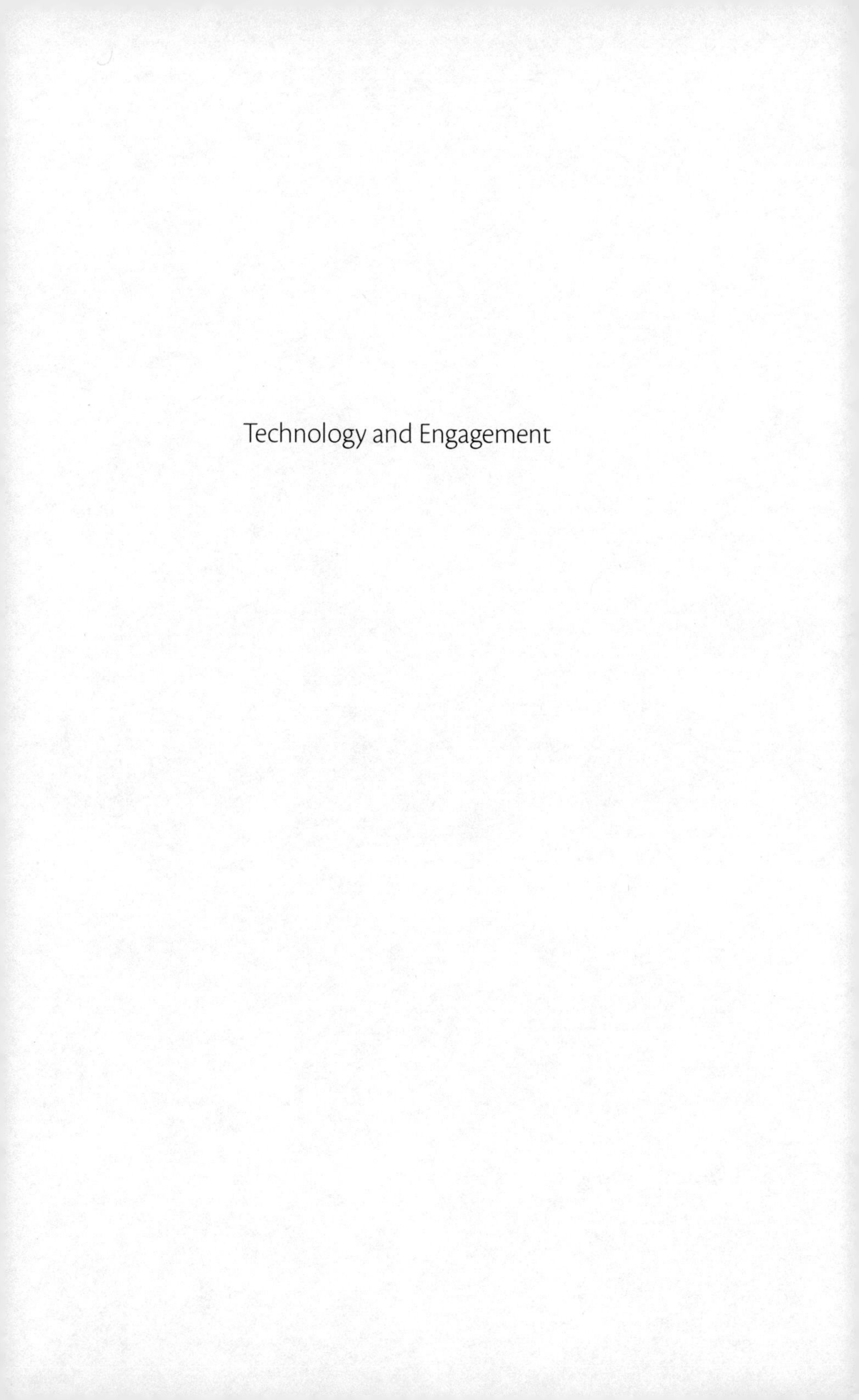

Technology and Engagement

Introduction

Yvette opened Facebook one Sunday evening to find photos of her family from a birthday party. Seeing all of her family and friends made her feel homesick. She started thinking about how much trouble she had been having in college lately. She thought about FaceTiming her mom, but figured she would wait until their regularly scheduled chat the following evening. A part of her didn't want to burden her mom right now with her feelings of uneasiness about being in college. Lately, Yvette had been wondering if coming to Millbrook University was the right decision after all.

Yvette was excited to attend Millbrook. As the first person in her family and one of the few from her neighborhood to go away to college, she arrived on campus ready to accomplish something no one in her family had. She knew the experience would be hard—the classes, meeting new people; however, she was also excited about the possibilities. In particular, she was excited to participate in a summer transition program prior to the fall specifically designed for students who were first in their family to attend college. When she began the program, she was surprised to find that so many of the students in the summer program at Millbrook were like her. Almost all of the students were Black, Asian, or Latino/a (like her). In many ways, the high level of support and camaraderie she experienced during that summer felt familiar, like what she had enjoyed in high school. Completing the program, she felt ready for the fall and her "real" college experience.

Once the fall came, and Yvette's summer network dispersed, her outlook began to change. Suddenly, being a student at Millbrook felt really hard. For starters, she found her coursework confusing and she often felt lost amid classroom discussions. She wasn't entirely surprised by this. It was no secret that her high school classes had not been challenging. Yet, she found herself lamenting about things she lacked—a computer to complete assignments, basic knowledge her peers seemed to possess, money to go out on weekends. But most surprising, and perhaps weighing on Yvette the most, was that she often found herself not knowing how things worked at Millbrook. Simple things, like the purpose of faculty office hours or what to do if the library didn't have the book she needed. At first, she thought this was typical of being in a new environment; however, Yvette noticed that many of her peers did seem to know these things and many other things that eluded her. She also noticed that unlike when she was in high school, she was reluctant to ask for help. It felt like asking for help at Millbrook would put a spotlight on her shortcomings and that seemed too risky in this environment.

Today Yvette and others like her are navigating difficult terrain as they transition from high school to and through higher education. First-generation college students (FGCS), those who are the first person in their family to enroll in college, are enrolling in higher education at higher rates than ever before. Armed with lofty aspirations and a desire to be the first person in their family to earn a college degree, often bringing pride and economic resources to their family, many FGCS begin college unprepared for the academic, financial, and social-emotional challenges awaiting them (Davis, 2010; Jehangir, 2010a). These challenges contribute to the gap between those students who stay the course and attain a college degree and those who give up, or worse, are forced to leave.

Growing concerns about this potential leak in the educational pipeline has the attention of higher education leaders and policymakers alike. Because access to a postsecondary degree in the United States has been identified as the great equalizer, apparent gaps in degree attainment between FGCS and their continuing education peers is one of the greatest challenges facing U.S. higher education.

Institutional leaders are well aware of the considerable challenges facing FGCS. Growing pressures for accountability have prompted university administrators to take steps to remove the obstacles that appear to hinder students' success. Importantly, practitioners have created a plethora of new transition programs, innovative student support models, and curricular reforms. These measures aim to address gaps in knowledge and skills and foster deeper engagement between students and their postsecondary experience. Yet we know that for today's young adults, social media and other forms of technology provide powerful platforms for interventions. However, to what extent have campuses considered social media as a mechanism for supporting FGCS completion of a college degree? We thought about this very question. We wondered whether and how students like Yvette might utilize social media and Web 2.0 technology, which facilitate user interaction, to access the types of campus information, support, and opportunities that are critical to success in college.

Access and Opportunity

As Baum, Ma, and Payea (2013) point out in *Education Pays 2013*, there is little question that earning a college degree yields multiple benefits to students and to the United States. The many benefits to students include economic stability through higher levels of employment, increased earnings, and the increased likelihood of receiving health insurance and pension benefits. Benefits of college also include healthy outcomes and passing educational benefits on to children. Ultimately, the benefits to society are equally as important. The nation benefits from higher levels of degree attainment through reduced health care and social service costs and increased levels of civic engagement and tax revenues (Baum, Ma, & Payea, 2013). These benefits are driving increased enrollment rates, federal policy, and shifts in higher education practice. Perhaps as a result of the possible gains, a spotlight has been placed on this postsecondary path. "College for all" philosophies permeate secondary education and society at large (Savitz-Romer & Bouffard, 2012). Increasingly, students whose parents have not attended college are aspiring to attain a postsecondary degree. However, when students pursue higher education, yet leave prior to earning a college degree, both direct financial costs and opportunity costs of time away from the workforce are significant. This type of attrition is costly to colleges and universities. Especially in light of growing accountability of institutions to both the government and students and families who have access to institutional data, attrition runs the risk of influencing their status.

FIRST-GENERATION COLLEGE STUDENTS ON CAMPUS TODAY

FGCS reflect one of the fastest-growing student groups in higher education. Currently, more than 50% of the parents of all 5- to 17-year-olds in the United States do not have a college degree (Aud et al., 2012). According to the National Center for Education Statistics, nearly one-third of all undergraduates are first in their family to attend college. Whereas enrollment among this student population is up, graduation rates have not followed suit. There is a notable gap in college completion rates between FGCS and continuing generation college students (CGCS). On average, less than one-quarter of all FGCS graduate with a bachelor's degree within six years of high school graduation compared to more than two-thirds of continuing generation students (Chen, 2005).

This trend depicting a growing population having access to, but not necessarily through, higher education is troubling. Against a backdrop of a nation determined to widen access to a postsecondary credential, one would hope that gaining access to college should lead to degree attainment. Yet, that is not the story we hear from many FGCS. It seems there are many factors working against FGCS postsecondary aspirations and plans. Academic, financial, social, and cultural factors create a transition to and through college marked with difficulty and barriers. The difficulties FGCS face during this transition can be attributed to many factors, including inadequate academic preparation, issues related to poverty or their parents' lack of postsecondary education, and the associated knowledge and expertise that are otherwise commonplace among continuing education parents (Davis, 2010; Jehangir, 2010a). In some cases, FGCS struggle with academic expectations that are not in line with their existing skills and knowledge. Making matters worse, such academic difficulties are often exacerbated by a lack of college knowledge or experience seeking out effective academic supports (Vuong, Brown-Welty, & Tracsz, 2010). For other students, it is the difficulty negotiating the campus culture and environment. Jehangir (2010a) suggested that FGCS lack a codebook, or a necessary point of reference, to guide their transition into a new context. Beyond academics and college knowledge, many FGCS carry jobs necessary to finance their higher education; however, this often leaves limited time for campus involvement or seeking help when needed. Together, these social, cultural, academic, and financial factors produce a constellation of challenges for FGCS as they make their way toward postsecondary degree attainment.

For many FGCS, these stressors can lead to departure from higher education (Inman & Mayes, 1999; Ostrove & Long, 2007; Warburton, Bugarin, & Nuñez, 2001). This departure is reflected in statistics showing FGCS earning lower first-semester GPAs than their continuing generation peers, leaving college after their first year, or taking longer to graduate (Chen, 2005; Engle & Tinto, 2008; Vuong, Brown-Welty, & Tracz, 2010). Even in cases where FGCS do not choose to leave higher education, their experience in college may not be a positive one (Davis, 2010).

THE ROLE OF HIGHER EDUCATION

There is a considerable body of literature describing FGCS students and how their transitions to college differ from those of their continuing education counterparts in college. Such studies have impelled institutions to initiate multilevel practices and programs to foster campus engagement and academic attainment among FGCS. These efforts seek to improve students' academic preparation, expand their information about and knowledge of college, create institutional support systems, and increase access to financial aid and scholarships. Support programs aim to address a range of needs, including logistical and transitional information, academic preparation and readiness, and social and relational support. These supports have been delivered through bridge and orientation programs focused on transition, ongoing student support services, and structural initiatives such as living and learning communities intended to integrate academic and social supports. Financing for these programs comes from federal dollars in the case of Federal TRIO Programs or directly from institutions themselves.

With the exception of structural initiatives, such as living and learning communities, the majority of existing supports have historically operated from a belief that FGCS, or other disadvantaged groups, arrive on campus with a deficit that needs to be remedied. This approach assumes that traditional institutional practices work for all students. With time and a shifting mindset about responsibility, more and more higher education institutions have come to see the importance of systemic change in service to better support FGCS, among others. New beliefs consider the contextual factors that might shape student engagement such as cultural norms, faculty expectations, institutional racism, and the presence of equitable access to supports. These shifts have brought about programs that target faculty development, structural changes such as learning communities, and curricular changes designed to invite students to co-construct knowledge with faculty.

The effects of these shifts are increasingly visible. For example, Estela Mara Bensimon and Alicia C. Dowd, at the University of California's Center for Urban Education (CUE), utilized CUE's Equity scorecard to work with multiple institutions to examine institutional policies and practices to produce equity in student outcomes. Their work on a Mathematics Department redesign at the Community College of Aurora included equity mentoring for faculty and a support lab for students enrolled in developmental courses. Over time students were more successful and there were more equitable outcomes across race and ethnicity (Center for Urban Education, 2016). Following suit, more institutions have begun to look at engagement through a critical lens, searching for specific institutional change levers. As we increase this examination of institutional practices, there is promise that we are moving closer to closing the gaps and promoting success for all students enrolling in higher education.

Whereas higher education leaders have taken many steps toward making their campuses inclusive and supportive for FGCS during their transitions to college, many scholars, including Museus (2014), have argued that there is an urgent need for "new tools and lines of inquiry" to better the chances for college success among these populations of students (p. 190). We believe that technology is one such tool that has yet to be fully leveraged to support the transition into and through higher education. Social media and mobile technology platforms are ubiquitous among college students. However, while usage among faculty is increasing, many are still reticent to fully take advantage of the opportunities such platforms may offer for alternate forms of cognitive and co-curricular engagement (Dahlstrom, 2015). Simply put, students are on social media and Web 2.0 platforms, and while they certainly use these platforms for entertainment and as sources of distraction, they also utilize them for important academic purposes. Such processes include, but are not limited to, using Facebook to connect with classmates to ask questions about homework assignments, forming online study groups, and having web-based videoconferences instead of face-to-face (F2F) meetings.

Through social networking sites such as Facebook and Twitter, users have access to information and knowledge that can serve to build weak ties or connections that cultivate inclusion, to establish social networks with new and different people, and to strengthen those bonds with people with whom they share norms and values. For college students, social media function as gateways to campus culture. Using smartphones, tablets, or laptops, college students employ social media to communicate, share, and connect with other students on campus (Martínez Alemán & Wartman, 2009). Transformative Web 2.0 communication technologies such as Facebook, Twitter, and Instagram engage the user as both a consumer and a producer of information. Users circulate social and cultural capital on these sites (Ellison, Steinfield, & Lampe, 2007; Ellison et al., 2014; Hofer & Aubert, 2013; Scott & Carrington, 2011). On college campuses, social media have been used to improve students' adjustment to college (DeAndrea et al., 2011) and to serve as a social lubricant that fosters the development of weak relational ties that provide students with new and unfamiliar information. It is through these emerging relationships, for example, that FGCS can request information and support (Ellison, Steinfield, & Lampe, 2011). Existing research, however, has not examined the value of FGCS relational ties for transition and engagement. For example, like all strong ties, FGCS bonds with family and home networks are likely consequential to their engagement and sense of belonging. But these enduring relationships can now be accessed more readily and more frequently through social media technologies. Do social media enable these relationships to serve particular functions in FGCS college transition and success? FGCS family and home networks are often highly motivated to support FGCS in their college transition in whatever way they are able. Do social media technologies extend their reach and impact?

Scholars have pioneered research on college students' use of Facebook and offer useful insight into the potential for social media to effectively mediate student engagement for FGCS. Junco (2011) examined the relationship between Facebook use and traditional benchmarks of student engagement, concluding that Facebook activities were strongly predictive of engagement. Brown, Wohn, and Ellison (2016) found that use of social media by FGCS provides multiple ways for students to learn about college life. Martínez Alemán and Wartman (2009) recommended that student leaders can act as peer-educators in helping new students develop best practices for impression management, a desire among new students who wish to manage how they are perceived by others. The work of these scholars, as well as others, suggests many promising possibilities for using social media and technology to promote engagement and access to campus capital.

Although campuses are finding new ways to utilize technology, the specific mechanisms through which social media and Web 2.0 technologies can be employed to enhance FGCS social and academic experiences on campuses have yet to be explored. Through our study, we investigated how institutions can use technology to provide FGCS the important supports necessary for their engagement and success on campus. To date, it appears that little has been done to study the role of technology in the transition and engagement of FGCS despite the fact that Web 2.0 technologies now occupy a dominant position in campus life. With this in mind, our project was firmly committed to understanding and engendering educational equity for FGCS through the use of social media and application software on tablet computers.

Knowing how to gain the information or knowledge to facilitate the college experience is critical for all students, but especially for first-generation college students. FGCS lack access to forms of social and cultural capital useful on campus (Walpole, 2003). Campus and college knowledge and resources can be understood as social capital, and their circulation on college campuses happens through networks of relationships or weak ties outside and inside the classroom. Specifically, various researchers have used social capital theory as a lens to understand the differing engagement and experience of FGCS.

Some scholars determined that FGCS experience a cultural mismatch between the college culture, which may reflect middle-/upper-class norms and independent expectations, and their own cultural norms, which may be rooted in working-class norms and value interdependence. Such a cultural mismatch occurs among FGCS, who are often racial and ethnic minorities, and typically have differential educational outcomes (as traditionally measured) (Pike & Kuh, 2005; Terenzini et al., 1996; Tym et al., 2004). Because they are the first person in their families to attend college, they are particularly vulnerable to those forces that limit access to the full range of cultural and social capital associated with college-going.

Awareness that FGCS might lack access to these important forms of capital has driven the growth of mentoring programs, bridge and transitional programs, and campus-based support programs, all of which aim to foster relational ties in service to the transmission of campus capital. What has garnered less attention, however, is how and to what extent FGCS preexisting relationships, or strong ties, offer important sources of capital for college success. Additionally, in what ways might FGCS weak ties, or newer connections that provide resources and information, offer important types of capital necessary for postsecondary success?

Current efforts to promote equitable opportunities to a college degree among FGCS have emphasized the importance of engaging students with their institutional campuses and cultures. These efforts are shaped by student engagement theories. Several researchers have concluded that active engagement in college promotes student persistence and ultimately graduation (Kuh et al., 2006). Student engagement reflects students' sense of belonging at an institution and has been associated with their utilization of resources, development of campus relationships, and ultimately success (Strayhorn, 2012). FGCS experience lower documented levels of engagement than students whose parents have completed college (Kuh et al., 2007). Many scholars have suggested that some students uniquely claim membership or belonging in the campus community. Specifically, they assert that research on college persistence inaccurately frames minorities' postsecondary experiences (Rendón, Jalomo, & Nora, 2000), thereby prompting institutional efforts largely charted by the research community's customary assumptions about student engagement and membership on campus.

Guiding Assertions

We hypothesized that conceptually campus capital could help explain how FGCS could (or do) access social capital on campus through networking and mobile technologies, whether these technologies could (or do) provide an effective means to that capital. Consequently, two assertions emerged that guided our project:

1. Web 2.0 technologies provide a unique avenue for FGCS to access campus capital through maintaining important weak and strong ties that are critical to students' success in higher education.
2. Social media provide a promising means of engagement for FGCS.

Project Background

This book is based on a five-year project that originated in the Educational Opportunity Program (EOP) at Millbrook University (institution and program names used throughout the book are pseudonyms). We chose this site based on prior relationships with program staff. Millbrook University is a highly selective, private Predominantly White Institution (PWI) with approximately 9,000 undergraduate students in an urban area of New England. Approximately one-third of all students who apply are admitted to Millbrook, and about one-quarter of these students decide to enroll. Admitted students have a composite range of SAT scores between 1960 and 2150. With a focus on the liberal arts, approximately two-thirds of students are enrolled in the College of Arts and Sciences. Close to 25% of students are enrolled in the College of Management, and fewer than 10% of students are enrolled in the Colleges of Education and Nursing. The top majors at the Millbrook include Economics, Finance, Biology, Communication, and Political Science. Thirty percent of students at the institution are students of color (SOC). Eleven percent of students identify as Asian; 12% identify as having two or more races/ethnicities; less than 5% identify as Black or African American; and under 3% identify as either Hispanic or Latino/a. More than 40% of undergraduate students received need-based financial aid; the average need-based scholarship or grant was more than $32,000, while on average a financial aid package was more than $37,000, including loans and work-study. When students commit to attending Millbrook, they tend to stay, with a 95% freshman to sophomore year retention rate and a graduation rate above 90%. Almost 100% of first-year students live on campus, and a majority of students live on campus for at least three years. Although FGCS at Millbrook graduate at rates higher than the national average, their graduation rates still lag behind those of their CGCS peers, and they do not always report a positive experience while at the institution.

Approximately 2,200 first-year students enroll at the institution each year. From this group, approximately 40 students are conditionally admitted through the EOP. Student participants in our project were selected to take part in this summer bridge program and were primarily underrepresented FGCS. At Millbrook, the goal of the EOP is to prepare a select group of diverse students who have shown leadership and potential in spite of challenging educational and financial circumstances to transition to Millbrook University. Through a six-week residential summer program, students are provided academic, social, and cultural support for success in college. Through residence on campus with peers, mentorship by peer mentors, enrollment in faculty-taught courses, and social activities, EOP students are given an orientation to the academic and social life of college prior to the arrival of the entire incoming class. Once students complete the program and start classes in the fall, they remain connected with their EOP advisor. Operating for more than 30 years, the EOP has provided important transitional support to Millbrook students. With more first-generation college students entering higher education and the continued diversification of college campuses, increased awareness about issues of access and equity have been raised. As a result, programs such as the EOP have traditionally been utilized to assist these students in attaining the capital required to successfully navigate the college environment.

Although it is not a requirement for selection, most of these students are FGCS, low-socioeconomic status (SES), and/or historically marginalized racially or ethnically. Students in our study were split evenly between males and females and were predominantly traditional age (97% were 19 years old or younger at the end of their first semester of college). Furthermore, 32.5% of the students did not speak English as their first language, and 90% of the students identified as non-White. Although half of EOP students estimated a family income under $30,000 per year, 84% estimated that their parental income was under $50,000 per year. When examining parental educational attainment, 60% of fathers and 55% of mothers had a high school diploma or less.

Our initial project utilized two primary components: iPad pedagogy and Facebook usage. Students who consented to participate in the research project were provided new iPads for their personal use over the course of the summer and through their first year. Students who stayed in the study after the first two years were allowed to keep their iPads. Students were encouraged to use their iPads during their EOP English and mathematics classes, and English and mathematics instructors were also given iPads to utilize for the summer when teaching their classes. The iPads were preloaded with academic and social media apps, which the research team, in consultation with the instructors, believed would be beneficial academically and socially. The Office of Multicultural Affairs (OMA), which coordinates the program, also created a Facebook group for these students, with the assistance of the research team, to promote social networking within the group. The goal was to create a seamless hybrid space for campus experiences often disconnected (whether in reality or perception)—a space where the academic and social are integrated, where faculty and students engage in public conversation and communication, and where peer processes can support the exchange of campus "know-how" and shared expectations for engagement. For example, on an iPad2, a student reading *Drown* by Junot Díaz can access grammar and syntax applications for reading response assignments, or academic-style applications to understand the nuances of paraphrasing and the dangers of plagiarism. Students and faculty can visit academic honesty sites simultaneously and work through the conventions of academic writing that are new to FGCS. Faculty can also make use of Kahn Academy coursework (http://www.khanacademy.org/) to supplement and enhance lessons. Faculty can iChat or Skype with students to review material or answer particular questions, or can direct students to particular assignments embedded in apps to build academic writing and analysis skills. Group work and discussions can take place easily on the group site, and students can use the technology for assignments (e.g., student group presentations on Prezi).

We integrated online social networking and tablet technology into the EOP to attempt to build an online culture of college belonging. We hypothesized that these technological capabilities could provide the pathways through which academic, advising, and mentoring experiences in the EOP's summer coursework could thrive. We also thought that the use of technology could facilitate relationships with faculty and EOP peer mentors that could continue throughout the academic year.

To arrive at an understanding of whether or how Web 2.0 technologies might serve as a useful tool for the transmission of campus capital, we collected data through focus groups, individual interviews, observations of Facebook usage, and classroom observations. These data collection methods included students, faculty, and staff across two cohorts with longitudinal follow-up. Detailed data collection information is provided in the Appendix.

In this book, we interrogate the important question of whether and how institutions can use technology to leverage campus capital in service to student engagement and success. As we listened to students, faculty, and staff associated with the EOP, we found that, in fact, social media and tablet technology can be used as a means to develop strong and weak ties that improve FGCS transition and engagement. These ties are connections between individuals that carry information through the social network, with tie strength determined by the intensity of the relationship (Granovetter, 1973). This is good news for institutions that are committed to ensuring equitable opportunities for success for all of their students, especially those most vulnerable to the challenges inherent in the postsecondary transition. Our experiences and findings suggest that embracing technology in intentional ways will indeed yield positive results.

We focus on first-generation college students and their use of Web 2.0 technologies as tools for academic and social transition to college. Though recent research has suggested that lower-income students with parents with lower levels of educational attainment do not use these sites to increase social capital and improve integration (Junco, 2013), our study concluded that FGCS do receive benefits through social media and additionally through application software on tablet devices. Our project revealed that FGCS use social media as part of an ecology of transition through which necessary strong relational ties with family and close friends at home are maintained. FGCS strong relational ties motivate, validate, and support students' identities, serving to secure their place on campus. Through the maintenance of strong relational ties, our FGCS were then able to move toward the development of weak relational ties and accessing newer forms of capital. In Chapter 1, we offer a hypothesized conceptual model to represent this phenomenon, and revisit the model in our concluding chapter. Consistent with Museus's (2014) "Culturally Engaging Campus Environments (CECE) Model of College Success" (p. 206), we present a conceptual model that is action-based and advances the latest technologies.

Organization of the Book

Recalling the frustrating experiences of our former students drew us to this project and is reflected in our methodological approach. For example, as former practitioners, and from some of our own experiences as FGCS, we understand that students' voices provide the clearest snapshot of the meaning that FGCS make of their experiences. As such, each chapter begins with a brief vignette, which illustrates many of the concepts and findings presented in the chapter. We expect that practitioners and researchers alike will resonate with the experiences described by the FGCS in our project.

Throughout the book, we aimed to incorporate the voices of the participants in our study as much as possible. Since we conducted more than 67 student individual interviews and 8 focus groups with close to 40 students and 20 individual interviews with 12 faculty/staff, and 4 focus groups of student staff, using pseudonyms for each of these participants did not seem practical. In order to simplify our story, we created 11 student composites based on the lives of our students. These composites are an amalgamation of several individuals in our study and are not meant to stereotype our study population. All of the direct quotes used in our study are actual quotes from our participants.

Here are the composite profiles of the 11 students in our study:

Andrew is White and from California. He is an Economics major with a minor in Spanish. He was an honors student at his predominantly Hispanic high school and captain of the hockey team. At Millbrook Andrew was selected to a competitive men's dance team. He sees his iPad as key in being able to keep up with his peers.

Arun is a Cambodian male who came to the United States when he was 6 years old. He lives in New Jersey—across the river from New York City. Arun is close with his sister, and connects with her daily on Snapchat or FaceTime to check in and see how things are at home. Arun and his sister are the primary translators for their parents, who speak little English. He is a Sociology major and a huge advocate of incorporating more technology into the classroom. Arun keeps to himself and is usually listening to music with his headphones on if he is not in class.

Ava is an African American female from Detroit. As the oldest of five children, she had a great deal of responsibility while growing up, protecting her siblings from the violence that was so prevalent in her community. She attended a charter school where one of her mentors was a Teach for America volunteer and alumnae of Millbrook University. She is often homesick and checks in with her mom and siblings daily—often on Snapchat or by text. Ava is a Political Science major who aspires to a career as an attorney.

Camila is an African American female from California who is unsure of her major. She loves the performing arts and was involved in dance and theater in high school. Camila has a large extended family, and her cousins are like brothers and sisters to her. She regularly stays in touch with them on Snapchat, Twitter, and Facebook. Just recently she joined GroupMe to be in contact with her EOP peers and people who live on her floor.

Carolina is a Latina Biology major who planned on going into pre-med when she started at Millbrook. She was always a top mathematics and science student at her New York City high school, but has really struggled with her science courses at Millbrook. She is beginning to think that Biology is not the best path, and has been loving her English classes. She connected with her EOP English instructor, who is a mentor to her on campus. Carolina switches between loving and hating Yik Yak and regularly

Snapchats with her cousins and high school friends. Carolina has joined a spoken word organization on campus and would like to intern somewhere during her time at Millbrook.

Gabriel is a Latino Psychology major from a smaller city in New England. He has always been an athlete; coming to college is the first time he has not been part of a varsity athletic team. While he has received a generous financial aid package from Millbrook, he needs to work multiple jobs to make ends meet since his family is not able to provide financial support. Gabriel depends on his iPad for class readings and taking notes in his classes, FaceTime with family and friends, and Netflix.

Krista is a Black female majoring in Sociology. Krista grew up as an only child and lived with her mom, and she feels like she needs to check in on her when she can. She usually FaceTimes or texts with her mom at least once a day. Krista is from a working-class town two hours away from Millbrook. She struggles to feel like Millbrook is the right "fit" for her, but found community by joining the Caribbean Culture Club and Black Student Forum. Over her time at Millbrook Krista has been active in #BlackLives-Matter protests on campus.

Mason is an African American male who lives in one of the poorest neighborhoods of the city where Millbrook is located. He was an honors student at one of the lowest-performing schools in the city. While Mason is attending Millbrook on full financial aid, he works as a security guard at night to help support his mom and siblings. He usually goes home at least once a week to make sure his siblings are all "on track" in school. He is part of a first-year emerging leaders program, and is interested in becoming an RA. Mason does not really like Facebook, but feels it is necessary to connect both on campus and with family and friends at home.

Meena is a female Indian American who attended a large, diverse high school in Maryland. She is an International Business major and plans to minor in French. She regularly uses Facebook to update her relatives about how college is going. Her passion is photography and she is always on Instagram. She just joined the Asian Christian Fellowship.

Nico is a Honduran male Finance major from a working-class suburb of Boston. He was the valedictorian of his low-resource public high school that was predominantly Black and Hispanic and did not have any AP classes. He often calls, texts, or Skypes with his parents for support during his challenging transition. He recently joined a men's mentoring group at Millbrook.

Yvette is a Latina Nursing major from Texas. She is close with her family and friends at home and is often homesick. She uses all types of social media to stay connected with her support system at home—FaceTime and texting with her mom, chatting with her Spanish-speaking grandmother on Facebook, and regularly connecting with high school friends on Instagram. Yvette was in the top 5% of her high school class and captain of the dance team.

Our faculty and staff composites are as follows:

Colin is a White Psychology major from Philadelphia entering his senior year at Millbrook. He is an EOP peer mentor and an alumnus of the EOP. He is planning on applying to the Master of Social Work program at Millbrook this year. He is an RA and works as a tutor in the Learning Center on campus. He is an avid Twitter user and has a GroupMe for the residents on his floor.

Lilly is a White female English instructor in the EOP. During the academic year she is the academic support coordinator for Millbrook's TRIO Program on campus. As a FGCS herself, Lilly has a large group of students who see her as a mentor on campus. Lilly was frustrated by the integration of iPads into the classroom during the first year, due to lack of training and time to prepare to incorporate this new technology. In the second year she incorporated multiple assignments using Facebook, which really engaged students with the literature they were reading.

Mark is Latino and the assistant director of the OMA. He coordinates the EOP each summer. Mark has been a staff member at Millbrook for the past 10 years. Students on campus, particularly men of color, see him as a resource and a support on campus. Mark does not use any social media because he does not want to let students into his personal life.

Martha is a White adjunct English professor who teaches in the EOP each summer. She has taught at Millbrook for more than 10 years and teaches multiple sections of English 101 each semester. She often ends up as the academic advisor of students in the EOP each year. She is flexible in her teaching style, and was excited about the prospect of using iPads in her classroom.

Ruby is an African American female from Boston and a peer mentor in the EOP. She is a junior Political Science major who plans on attending law school after graduation. Ruby is involved on campus and was recently elected as president of the Black Student Forum. She regularly Snapchats with her sister and has had more than 200 up-votes for her posts on Yik Yak.

Throughout the book, the following acronyms are used:

CGCS	continuing generation college students; students who will not be first in their families to graduate from college
EOP	Educational Opportunity Program; pseudonym for the campus bridge program
F2F	face-to-face communication
FGCS	first-generation college students
LIFG	low-income first-generation
NSSE	National Survey of Student Engagement
OMA	Office of Multicultural Affairs; pseudonym for the campus office working with racial/ethnic minority students
PWI	Predominantly White Institution
SES	socioeconomic status
SOC	students of color
STEM	Science, Technology, Engineering, and Math
TA	teaching assistant

ROADMAP OF CHAPTERS

In the following chapters, we present the case for Web 2.0 technology as a means for FGCS to better transition and engage in college, with emphasis on the empirical and conceptual guide for the project. By summarizing the current social science knowledge that motivated and inspired our project and this book, we aim to provide the reader with an empirical and conceptual foundation necessary for understanding the findings from our project. Throughout, we present the central findings of our five-year-long project with a focus on students' social and academic experiences. Though many scholars have used social capital theory to understand students' experiences in college, in Chapter 1 ("Engagement and Campus Capital") we present the conceptual framework for our study. In Chapter 2 ("Being First-Gen on Campus"), we describe the social science literature on the experiences of FGCS on campus, in particular, those who are students of color, those who are low-income and attend PWIs to scaffold our findings. In this chapter, we share many of the concerns expressed by the FGCS in our study, calling attention to the constellation of issues that shape students' early transition to college. Chapter 3 ("Web 2.0 Technologies on Campus") introduces the reader to the advancement of Web 2.0 technologies, in particular, social media and tablet technology, and the landscape of these media on the college campus. Through students' reports of social media usage, we make clear in this chapter the value of these technologies for improving access to campus capital for FGCS.

Since the study's main interest was to understand the ways in which technology could be used to provide FGCS access to capital in students' college experience, we were reluctant to categorize their experiences too narrowly. However, we chose to disaggregate academic and social experiences in service of greater clarity. Consequently, we examine how particular types of social and cultural capital on a college campus are critical for FGCS transition and engagement in Chapter 4 ("Transition and Campus Engagement"), and we present in Chapter 5 ("Bridges to Campus Capital in the Classroom") the voices of FGCS and their faculty as we look at the ways in which social media and tablet applications affected their academic transition and engagement. In these chapters, student accounts depict how capital is particularly important for FGCS and their integration to campus life.

The book's concluding chapter, Chapter 6 ("Propositions for Change"), offers a discussion and presentation of programmatic change recommendations informed by our project. Specifically, we detail how campus administrators might leverage such technology to encourage student success, presenting our revised conceptual model of the role of technology for supporting FGCS transition and success. In sum, this book offers a summary and synthesis of what is known about FGCS transition and offers a new conceptual framework and rationale for how institutions can make technology work for FGCS.

THE PURPOSE BEHIND THE BOOK

Our aim in writing this book is to present the conclusions drawn from our project with the hopes that faculty and administrators responsible for advancing the success of first-generation college students will consider ways in which Web 2.0 technology can be a lever of equity. A growing number of colleges have dedicated programming toward FGCS; thus there are many places where recommendations can be used in practice. Specifically, we expect that faculty and administrators at schools like Millbrook University, as well as the many other institutions committed to supporting the engagement and achievement of FGCS, will find practical ideas and lessons drawn from our study. Administrators of Federal TRIO Programs, student support, and institutional support programs will gain an understanding of how FGCS experience the transition to college and specifically the potential for technology to ease that process. We also anticipate that the lessons learned from our study will have relevance to anyone who supports equity and opportunity in higher education. For example, there is a growing trend among nonprofit and community organizations to extend traditional high school support programs to the transition to and through higher education. In this way, we expect that organizations working outside higher education will gain an understanding of FGCS experiences with institutional initiatives during their transition. Finally, given the scope of this issue across higher education, we expect that policymakers concerned about the high school to college transition will discover that our findings have implications for new higher education policy as well.

We hope that readers take away from this book an understanding of how FGCS can access capital through technology to foster the transition to college and engagement on college campuses. We see technology as offering a uniquely powerful potential for the transmission of important capital without which FGCS are at a significant disadvantage. However, as the students in this study tell us, they experience their place in higher education through their identity as first-generation students and thus they are managing membership in multiple social spheres simultaneously. As such, and as we will explain in this book, embarking on technology-infused initiatives should be done with care and attention to FGCS multiple contexts and priorities.

1
Engagement and Campus Capital

When Carolina started at Millbrook, she was nervous about taking chemistry. She knew she needed to take it to meet her pre-med degree requirements, but she was nervous about passing it. Her high school chemistry class wasn't hard and she had a feeling that was because she didn't learn much. After a few weeks, Carolina grew to really like her chemistry professor, Dr. Arroyo. He was an engaging professor and really seemed to take an interest in his students. However, although she really enjoyed Dr. Arroyo as a person, she found his class difficult. After not doing well on a few tests, Carolina made an appointment to speak with him about her struggles with the content. It was then, at his office hours, that Carolina learned that Dr. Arroyo was also the first person in his family to go to college and even got his PhD! Carolina was especially surprised to learn that, like her, Dr. Arroyo was originally from New York City. In fact, Carolina knew the high school he had attended because her cousin went there. Carolina really appreciated knowing that she had a lot in common with one of her professors.

During this first meeting, and several that followed, Carolina described not being able to handle the work in his class. At first, Dr. Arroyo tried to give Carolina suggestions for studying and learning the new content. He introduced her to an upper-class Chemistry major to give her additional tutoring; however, it didn't seem to help. She continued to feel confused and lost in the course. At one point, Dr. Arroyo started to ask questions about why Carolina had picked pre-med as her major. The more she tried to answer him, the more she and Dr. Arroyo realized that perhaps she should consider some other options. Over the next few months,

Dr. Arroyo helped Carolina transition from pre-med to English. What a difference! Before too long Carolina was loving her classes and earning strong marks in all of her courses. She felt much better, and even though she earned a C- in chemistry she still considered Dr. Arroyo a mentor.

First-generation college students (FGCS) like Carolina often struggle academically and are more likely to withdraw from college before completing their degrees at rates higher than those of their peers with college-educated parents. We know that in order to improve their persistence, colleges and universities should provide FGCS opportunities to acquire the academic capital that positively affects persistence (Chen, 2005; Engle & Tinto, 2008; St. John, Hu, & Fisher, 2010). Researchers have identified processes that can develop academic capital, processes that include acquiring knowledge about navigating campus and the development of informational and supportive relational networks and behaviors (St. John, Hu, & Fisher, 2010). Structural factors like lack of access to mentors have also been identified as key concerns in improving FGCS and working-class student postsecondary persistence (Stephens et al., 2015). Relationships and communication with parents, teachers, and counselors are deemed highly supportive and influential for FGCS and low-socioeconomic status (SES) college students, and yet these are students who typically do not have the information or social capital to provide instrumental support to FGCS (Sy et al., 2011). These structural factors support and reproduce dominant group norms that are often outside the awareness of FGCS. These norms are prized and assigned value in educational settings and are understood as social capital. Non-dominant and socially marginalized groups in educational settings are frequently challenged to take on these norms, often as a result of limited information, incomplete knowledge, and/or lack of access to individuals who hold such capital (Stanton-Salazar, 1997).

As mentioned in the Introduction, colleges and universities engage in interventions designed to improve the persistence rates of first-generation college students by focusing on the logistical challenges of navigating college that these students face. For example, institutional practices and programs such as learning communities and peer-mentoring programs aim to communicate information and develop FGCS awareness of the many practices and procedures common on campus, but normally focus on developing academic and study skills (Bettinger & Long, 2005). However, research on the effectiveness of these conventional programs suggests that they can be ineffective (Charles A. Dana Center, 2012). Historically, these interventions have employed conventional face-to-face (F2F) strategies that do not take advantage of up-to-date Web 2.0 applications and mobile technologies, conceivably limiting FGCS ability to engage with the twenty-first-century college campus.

In many colleges and universities, programs to address FGCS transition to college, in particular, those aimed at improving their academic preparation and engagement, attempt to transmit social capital through instructor, staff, and peer mentorship relationships. In these varied programs, the primary goal is most often to improve FGCS persistence and success in college by increasing the relevant social capital of FGCS through F2F relationships. Courses and workshops on study skills and academic writing and on navigating curricular requirements are the centerpieces of most of these programs. Colleges and universities rightly have looked to relationships to transmit social capital to FGCS, but have done so through traditional means. What about the new relational spaces that now characterize our campuses? For example, as a new relational space can social media serve to transmit social capital? Early research on Facebook and social capital transmission showed a strong association between Facebook use as a means to obtain social capital, especially social capital or information that benefits the user in some way (Ellison, Steinfeld, & Lampe, 2007). Are social media another means to access the forms of capital that FGCS critically need to achieve academic success on campus? Could Carolina's use of Facebook enable access to resources or information like she obtained from Dr. Arroyo?

The experiential and cultural backgrounds of FGCS are most often disparate from those of continuing generation college students (CGCS). FGCS are often racial and ethnic minorities, usually come from low-income families (Chen, 2005), and typically have differential educational outcomes (as traditionally measured) (Pike & Kuh, 2005; Terenzini et al., 1996; Tym et al., 2004). The funds of knowledge FGCS bring to college reflect their non-college-going familial backgrounds and those social and economic forces that limit access to the full range of social capital traditionally associated with effective college-going. Often, there is not an obvious and precise correspondence between FGCS funds of knowledge, or cultural capital (Moll et al., 1992), and the social capital necessary for efficacious college-going. Consequently, as Dewey (1915) observed, educational institutions should seek the "keys which will unlock to the [student] the wealth of social capital which lies beyond the possible range of his limited individual experience" (p. 104).

In the twenty-first century, is technology the key to unlock the wealth of social capital beyond FGCS experience? Can tablet technology (like the iPad) and social media prove effective keys to unlock the means to access social capital on campus that is particularly relevant to FGCS? Early research on Internet communication suggested that users could gain emotional and instrumental support communicating on online networks (Wellman et al., 2003). Through connection strategies (Ellison, Steinfeld, & Lampe, 2011), social media may be a mechanism that colleges and universities could use to enable FGCS to build a network of capital-rich relationships. Current research indicates that social media are essential for a group of people to communicate purposes, to provide guidance and support for new members, to supply members with useful information and opportunities for personal growth and development, and to serve as a conduit for interaction between and among members so that they actively construct the community's culture (Martínez Alemán & Wartman, 2009; Nagele, 2005). Tapping their networked relationships for information and social and emotional counsel on social media sites like Facebook can positively affect self-esteem among lower-self-esteem users (Steinfield, Ellison, & Lampe, 2008). Facebook is understood as a site in which users draw on their network relationships for social capital (Brooks et al., 2014). Facebook users can derive more value from their online connections when they recognize that these relationships are resources of valued social capital (Burke, Kraut, & Marlow, 2011).

But how are we to understand technology and social media as keys to expanding FGCS access to social capital on campus? How are the distinctive capacities of tablet technology and social media well suited to address the unique social capital needs of FGCS? To answer these questions, one must consider the conceptual basis for these inquiries. This chapter examines what we mean by social capital, why access to social capital is so critical for FGCS, and why new technologies are specifically intended to extend users' social capital reach.

Our Conceptual Framework: Campus Capital

Threading together knowledge of FGCS and student engagement, Web 2.0 technologies, and social capital acquisition theories, our research project proposed a conceptual framework, "campus capital" (Figure 1.1). An inclusive term that consists of the various forms of social capital that enhance students' on-campus experiences that researchers have documented affect their persistence to graduation (Horvat, 2000; Pascarella et al., 2004; Walpole, 2003) through relational networks, campus capital enabled us to view and examine the various ways in which FGCS access social capital through Web 2.0 technologies. We define "campus capital" as those forms of explicit and implicit social capital specific to a particular campus. In this way, campus capital is derived from a framing of FGCS experiential reality through critical standpoints and theories of social capital acquisition, and the relationship between online social networking and social capital acquisition.

Our conceptualization of campus capital is informed by critical theorists, especially those focused on social capital acquisition. Critical theorists have sought to challenge the many different ways that individuals and groups are suppressed in communities. Horkheimer (1982) intended critical theory as a way to explain

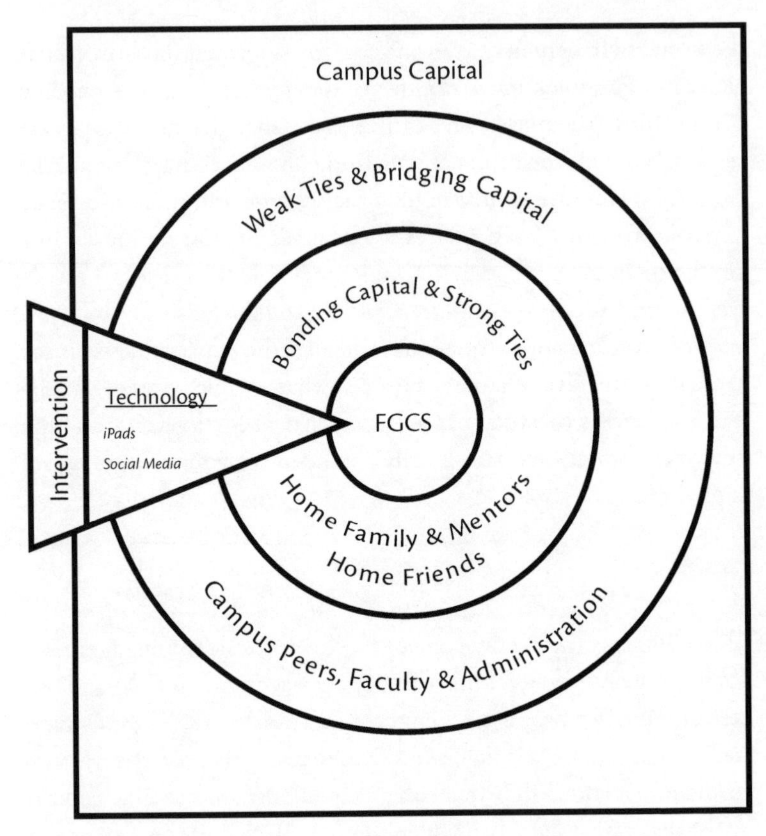

FIGURE 1.1. Proposed Campus Capital Conceptual Framework

the conditions in which an individual's agency was diminished by social structures. Foundationally, critical theory was aimed at offering options to extend and expand individual agency in all manner of social existence. Historically concerned with the ways in which capitalism as a social structure decreased individual agency, critical theory nonetheless evolved to enlarge its view of social inequality through Habermas's (1981, 1987) consideration of the role of symbols and language in determining social inequality and Levi-Strauss's (1963) critical anthropology that put forth a view of culture as a system of symbolic communication that circulates in institutions to produce patterns of behavior, values, and traditions that reinforce hierarchies and restrict agency. But it is

Bourdieu's (Bourdieu & Passeron, 1977; Bourdieu,1986) formulation of communication and culture as regulatory that has enabled us to identify the many ways in which colleges and universities are institutions in which FGCS access to campus capital is regulated through institutional culture, and the ways in which online social networks can aid in the acquisition of campus capital.

SOCIAL CAPITAL IN THE CAMPUS CONTEXT

From Bourdieu's (Bourdieu & Passeron, 1977; Bourdieu, 1986) critical perspective, a college or university is a multifaceted social space occupied by students, faculty, and staff, all individuals who communicate through a complex network of relationships. As individuals occupying this social space, students transmit knowledge and information through their network of relationships on campus. Through language (symbolic or otherwise) and behavior (action or inaction), individual students produce, transmit, and consume information, or, in Bourdieu's lexicon, capital. Capital is communicated through an individual's network of relationships on campus and its value is dependent on its utility in the social space, the campus. Students can transmit capital that Bourdieu would characterize as cultural capital or insider (tacit) knowledge that is derived from a student's membership in a particular class or group. In the case of college students, cultural capital can take the form of tacit knowledge that is the result of their own personal or family's social/economic class status, and/or their experiences with college culture prior to attending college themselves. Students may have college-educated parents who transmit cultural capital to them. This capital can include knowledge about living in the residence halls, attending class, choosing a major, or how to speak to and interact with faculty. College students may have siblings with college experience who also transmit cultural capital about college through the family's relational network. Students themselves may have first-hand experience of college because they attended a university's summer program while in high school, and/or may have received comparable cultural capital living independently at sleep-away camps or in high school exchange programs. Further, the contacts and memberships that students have before they enter college and those that they form while in college are sources of a form of cultural capital, social capital, that can be transmitted and leveraged in college.

Bourdieu's claim that forms of capital are regulated by educational institutions through norms, customary conduct, and activities (Bourdieu & Passeron, 1977) suggests that on campus individual students and groups of students may be experientially and structurally positioned outside the networks of relationships through which campus capital is transmitted and/or shared. Insiders to the culture of higher education and the culture of a particular campus possess, transmit, and leverage campus capital, effectively exercising their autonomy and agency because they possess that capital. Possessing campus capital enables individual agency. With campus capital, students can act autonomously and are not dependent on or subjugated by those students with capital. Outsiders to prevailing college culture may lack campus capital that is normative on campus, though they may possess different social knowledge and forms of capital. But more importantly, their agency on campus is diminished if the lack of campus capital restricts resource acquisition. Like Bourdieu's characterization of socioeconomic class as a matrix of resource inequality, students who do not have access to the full range of campus capital and its circulation find their agency constrained. Their decisions, actions, and movement in the social space are inhibited. Their engagement with the campus is constrained.

As Davis (2010) observed, FGCS are "unfamiliar with the culture of college," meaning the "insider knowledge, the special language, and the subtle verbal and nonverbal signals that, after one has mastered them, make one a member of any in-group, community, or subculture" (p. 29). Possessing insider knowledge gives one membership in the campus culture. Many institutions dedicate programs and services to familiarize FGCS with processes like registration and advising, but authors like Davis posit that FGCS "must also master a more profound set of skills and behaviors" that constitute the "complicated and subtle" aspects of campus culture (p. 30). Consequently, FGCS are inhibited by their limited access to resources embedded in the campus culture. Access to embedded capital is what determines individual agency and self-sufficiency, and that enables students to gain comfort and membership in the college community. A student's access to embedded capital on campus is "an investment in social relations with expected returns" (Lin, 2001a, p. 6).

Though Davis (2010) implied that relationships and networks are important to the social capital of campus life, we use the term "campus capital" to foreground the importance and significance of social networks for college engagement, the process through which insider knowledge is acquired. Campus capital is the engrained resources in the social relations and networks of students, faculty, and staff on campus that if utilized can enrich student life. As Lin (2001b) observed, the circulation of information across social networks provides individuals with useful information about opportunities and alternatives (insider knowledge) that as outsiders they do not possess. FGCS outsider status in college cultures confirms their lack of social credentials and social capital. Additionally, as in any culture, FGCS outsider identities will be reinforced the longer they are restricted from accessing insider capital through relational networks.

Campus capital is insider knowledge and information garnered through networks that circulates and is communicated through these relational networks. Campus capital is characterized by Lin's (2001b) elements of social capital: it is information attained through networks and gives individuals access to opportunities or decision-making that will impact their ability to claim membership in a community and that will enable them to be entitled to certain resources and community membership status. A comprehensive construct, campus capital is that network-bound knowledge and information that can "approximate being oriented intuitively" on campus (Davis, 2010, p. 33) and that serves to enhance students' on-campus experiences and affect their persistence to graduation. As Davis (2010) correctly upheld, unlike continuing generation college students, first-generation college students do not receive insight about college-going and college culture through their close, trusted, parental, and familial relationships. Though certainly supported and motivated by these relationships, FGCS development does not cultivate an intuitive awareness that positions them within or inside college culture. Campus capital extends understanding of insider knowledge and social capital to focus on the networks of campus relationships that act as a means to obtain insider knowledge or social capital.

This intuitive awareness enables CGCS as insiders to access varied forms of capital inside and outside the classroom. Within the classroom, FGCS often feel like imposters (Hayes, 1997). Positioned as outsiders to the normative culture of the classroom by virtue of their academic under-preparation (Pascarella et al., 2004) and their lack of intuitive awareness of college classroom etiquette and protocol, FGCS are often anxious about professors' and peers' expectations. How and when to contribute to class discussion, and when and whether to volunteer a response or comment are behaviors that are not informed by insider capital but rather by FGCS outsider intuition. The value of argumentation for the purpose of knowledge-building may seem foreign given FGCS academic experiences in high school, as can the culture of intellectual risk-taking common in many college classrooms. The risk of being incorrect coupled with not being confident in the style of normative college classroom communication may silence FGCS (Davis, 2010). The forms and customs of expected student behavior that signal engagement in the normative culture of the college classroom, and the jargon and lexicon of both disciplinary and academic culture, is most likely unfamiliar to FGCS. FGCS may struggle in class and their success in a course may be affected by their unfamiliarity with the lexicon of academia or academic language, especially if they are English-language learners (Scarcella, 2003). Could social media like Facebook that encourage students' collaboration and cooperation and enhance their social presences (Cheung, Chui, & Lee, 2011) prove effective for FGCS in the college classroom?

Consequently, just as in any club or organization, membership comes with certain rights, power, and privileges; however, those who do not belong are not afforded full membership, and as a consequence their range of resources and funds of knowledge can be regulated and inequitable. In a racist and oppressive society, membership in educational communities is restricted (in part) because of the regulation of cultural capital broadly understood (Bourdieu & Passeron, 1977; DiMaggio, 1982; Harker, 1990; Ladson-Billings & Tate, 1995). Consequently, membership in an educational community necessitates that members engage by accessing the community's insider knowledge and its networks of social associations and connections. Accessing the community's capital enables membership that enhances community engagement and an individual's sense of belonging. Being a member of a community requires the means to access the community's varied forms of capital and engagement with those processes through which social capital is expended.

In educational institutions like colleges and universities, individuals and groups have been historically excluded from full membership in the campus community. For example, though Oberlin College was the first co-educational college, women were not allowed to take certain courses. Women had to enroll in the Ladies Course that prepared them for marriage and motherhood (Linder, 2008). At Harvard College, African American men were not allowed in the residences until 1923; after they were allowed, they were segregated until World War II (Bailyn et al., 1995). By virtue of their cultural positions, certain individuals and groups are prevented from being fully engaged and from experiencing a deep and robust sense of belonging on campus. The regulation of race, ethnicity, class, and FGCS identity positions in campus cultures can determine the extent to which these historically marginalized populations have access to knowledge and its consequent benefits on campus and off (Bourdieu & Passeron, 1997; Yosso, 2005). Social regulation dictates and defines the ways in which students do and do not belong on campus through discursive norms around such things as academic competency, academic communication, and social networking (Walpole, 2003). For example, colleges and universities regulate students' membership in the campus community through discursive assumptions about students' knowledge of postsecondary norms and values and their correlate processes. It is often assumed that all students are familiar with the tacit knowledge of academic and social norms and expectations on campus. Knowing these social and academic norms is a condition of membership and contributes to students' success (Pascarella et al., 2004; Walpole, 2003).

Accordingly, colleges and universities should provide the means through which students can acquire campus capital to help develop a sense of belonging and through which membership can be achieved (Hurtado & Carter, 1997). Like other educational institutions, colleges and universities are responsible for the access to membership and the extent to which individuals believe that they belong on campus (Johnson et al., 2007). These are institutions in which there exist many means of acquiring campus capital that prove to be fruitful investments for life on campus and life beyond college (Walpole, 2003).

For social media to serve as the means to transmit campus capital, they must enable the production, circulation, and consumption of information across insider and outsider relationships. These technologies must enable FGCS to consume insider campus capital, as well as circulate and consume outsider campus capital. Skills, information, dispositions, and tacit knowledge that have enabled FGCS to succeed in high school and to gain access to and enroll in college can be understood as outsider campus capital. As outsiders to the norms of college-going culture, FGCS nonetheless bring culturally specific behaviors, knowledge, and attributes that can have value for their success on campus. For example, FGCS students have learned some level of self-regulation in order to achieve in high school and to enroll in college. This self-knowledge, the relationship between learning and effort, is an example of cultural capital that within FGCS communities can be transmitted in the familiar language and discourse of socioeconomic class through social media. As Savitz-Romer and Bouffard (2012) pointed out, the development of a "college-going identity" (p. 64) among FGCS and other youth requires the building of cultural and social capital in the middle and high school years through intentional schooling strategies aimed at identity development. These intentional approaches build self-efficacy in FGCS that is context specific and that is motivated by the unique socioeconomic and cultural context of their upbringing and development, as well as their schooling experiences. This capital can be circulated on social media as outsider campus capital across certain social networks.

Insider campus capital is the information and knowledge that is normative in college cultures. As a normative culture, it is highly valued and its historical context is informed by race, ethnicity, and class assumptions. Knowledge and information about academic comportment and expectations are characteristic and often inferred or implied in communication on campus. For example, in her review of community college students, Cox (2011) found that FGCS were unlikely to utilize faculty office hours for academic support and guidance. Recognizing the purpose and value of attending a professor's office hours, or even appropriate ways of engaging with faculty in this context, is one type of campus capital that may not have been procured by FGCS. As reported in recent qualitative research, this type of instrumental campus capital can be procured through social media during students' first-year transitions and positively impact their sense of belonging on campus (Vaccaro et al., 2015).

SOCIAL NETWORKING THEORIES AND CAMPUS CAPITAL

Social capital is a product of both an individual's network of relationships and the network itself (Putnam, 2000). It is in and through our social networks that capital (tangible and intangible resources) is exchanged and circulated. In our social networks, we perceive the acquisition of capital to varying degrees, but we are aware that relationships in our social network serve many purposes and affect us in a variety of ways (Brooks et al., 2014; Putnam, 2000). Bourdieu (1986), Coleman (1988), Putnam (2000), and Lin (1999, 2001a, 2001b) described the nature of social capital within social networks as layered, reflecting our situated and experiential contexts. These theorists asserted that, whether at the group or individual level, social capital transmission in networks involves reproduction of group norms, commonality and cohesion between group members, advancing and participating in identification with others in the network, and some degree of investment in the relationships that define the network. Most importantly, social networks provide the opportunity for acquiring resources that are intrinsic to the networks.

Because social capital is both the social network and its effects, it is important to understand the constitution and basis of networks, the expectations that we have of our networks, and the properties and outcomes of the interactions within the network. Social networks, whether F2F or online, are composed of binary relationships between individuals or relationships between and among group members. Networks vary in density. Networks in which people have actual ties with the majority of possible ties (e.g., families) are dense networks. Networks in which actual ties are far fewer than all of the ties possible (e.g., ties only to roommates and not other students in the residential unit) are less dense. The density of social networks indicates the strength of relational ties that compose them (Scott, 2012). Dense, cohesive networks give access to homophilous others who understand our experiential background and with whom we can share emotional intensity (McPherson, Smith-Lovin, & McCook, 2001).

Relationships in social networks have been conceptualized by Granovetter (1973) as ties, the potency or strength of which is determined by a combination of "emotional intensity," "intimacy (mutual confiding)" (p. 17), the length of the relationship, and the kind of reciprocity and social capital exchanged. The types of relationships that foster the transfer and acquisition of social capital are often referred to as weak and strong ties, and are a key component of social capital acquisition. These different ties are associated with different strengths shaped by intensity, intimacy, reciprocity, and time and that determine their social capital currency (Granovetter, 1973). Tie strength is also influenced by level of education, race, gender, SES, and other social class identifications and designations (Lin, Ensel, & Vaughn, 1981). Who we are as sociocultural identities has effects on the nature of our networks and the ties that compose them, thereby determining the information, norms, and expectations to which we have access. However, the variety of embedded resources in a social network—the diversity of the network—also impacts social capital acquisition.

WEAK TIES AND BRIDGING CAPITAL, STRONG TIES AND BONDING CAPITAL

Weak ties and strong ties, then, constitute social networks. The sociology of networks suggests that ties between users are either weak or strong and that the distinction between these determines the value of the communication (Granovetter, 1983). The distinction between weak and strong ties determines its capital and its intended and unintended effects. Weak ties and strong ties are interpersonal interactions or connections that carry information through social networks. The type of social tie considerably determines social capital circulated on social networks. It should be understood, however, that social networks are not static nor are the ties that compose them. At times, social ties in networks change and fluctuate, corresponding to the evolutionary character of our sociality. Consequently, the distinction between strong and weak ties is often vague and transitory. Easley and Kleinberg's (2010) example of the changing and subtle nature of strong and weak ties among teammates in-season and off-season exemplifies this point well. During the athletic season, teammates establish strong ties due to their shared norms and objectives and the need of each other's capital to attain those objectives. Off-season, ties among many teammates shift to weak ties because they are no longer exclusively bound by their similar context, norms, and shared objectives.

Weak ties are with acquaintances not proximate in social networks; they are not located in clusters of close proximity and do not compose cliques that specifically connect several users (Watts, 2004). Weak ties have strength in that they are greater in number and provide a broader and wider range of information. Because weak ties are more diverse, more unlike us, through them we have the potential for greater social capital gain (Granovetter, 1973). Through weak ties, we have access to new information, or more specifically, information that is unfamiliar or unknown within our relational (social and cultural) ken or our consciousness. Hansen (1999), for example, demonstrated that knowledge is transferred effectively and efficiently through weak ties among acquaintances at work, among colleagues with no close connections. The knowledge that is transmitted across workers' weak ties is information that is expedient and practical for the task at hand, making it capital. It is information that gives knowledge necessary for task completion, and in other ways connects and is a bridge to unfamiliar funds of knowledge. As Carolan and Natriello (2006) pointed out, weak ties in learning situations can determine the extent to which students (in the case of social media users) are deprived of information if their networks are composed of few weak ties, or if their networks are densely or exclusively populated by strong ties and cut off from wide-ranging weak ties. Students with few weak ties have only the perspectives and information from close, strong ties (those people most like them) and are unlikely to easily access valuable information outside those ties. The insularity of strong ties, though rich in culturally specific information/capital, restricts the user's access to a broader, more diverse range of capital.

Knowledge or information that we obtain through weak ties is characterized as bridging capital (Putnam, 2000). Bridging capital is social capital we acquire through ties with heterogeneous network members across weak ties, but more importantly, it is capital (information) that will give the user/consumer an advantage (Granovetter, 1973). Bridging capital is acquired in communication through weak ties in which trust is thin or not as resonant as trust between members in strong-tie networks (Anheier & Kendall, 2002). Our weak ties link us to social capital (resources and knowledge) that we do not currently hold but that has social value. Consequently, the value of weak ties is greatly determined by the bridging capital's utility (Lin, 2001a) and the extent to which we have interactions with others with diverse or different views and their correlated capital (Williams, 2006). If we perceive the social capital resident in our weak ties as having little value, if the capital is unrelated to our needs, or if we do not accurately perceive the capital value, then those ties can be ignored, underutilized, disregarded, or simply overlooked. In effect, the bridging capital is rendered ineffective.

The capital FGCS can access through weak ties is not likely to be found in the strong ties that they have with trusted, inner network relations most like them, whether they are high school friends, siblings, or other FGCS on campus. It is capital that is either unfamiliar or novel, or perhaps temporarily incomprehensible and needs translation. Because bridging capital is distributed through weak ties in the dominant college network, it is capital constructed within a college-normative cultural frame and is likely foreign to FGCS. Whether a consequence of lexicon or cultural discourse, such capital may need cultural translation. Martínez Alemán (1999) described this type of interpretation as "the language of critical translation" (p. 38). Davis (2010) identified this in the network of weak ties in a college classroom and as the interpretation by an insider guide.

In social networks, strong ties are with trusted others with whom we share embedded similarity. Homophily characterizes these ties. These are network ties that are regulated by experiential context, by our social and cultural correspondence. These are ties with friends and relatives whose own social networks overlap with our own. Loyalty and obligation reciprocity are typical of these ties, and established levels of intimacy and trust inform them. We have durable personal connection with our strong ties (Granovetter, 1973; Scott, 2012). Accordingly, these are ties that are readily available to us; the actors in strong ties are highly motivated to attend to our emotional and psychological needs. Emotional intensity and intimacy are essential mechanisms of strong ties that reinforce their stability and endurance. This affective character of strong ties makes them especially useful when we are experiencing self-doubt, indecision, apprehension, insecurity, or uncertainty (Granovetter, 1983).

As a consequence of homophily, the social capital that we access from strong ties is often superfluous and familiar. A network of strong ties is not "a channel for innovation" (Krackhardt,

1992, p. 26) because the information or knowledge that we acquire through these ties is often not the kind of knowledge that presents us with new ideas or options or with creative suggestions. It is often knowledge we already possess. The currency of strong ties is much more psychological and affective; it is not a currency of instrumental support. A network of strong ties is most valued for its ability to motivate us, to boost our confidence, and to extend sympathy and empathy. Because we share the same or similar cultural capital with our strong ties, we often turn to them for emotional and psychological sanctuary and shelter from challenging and trying times in our lives. The psychological and emotional support that we obtain from strong ties is bonding capital (Putnam, 2000).

Bonding capital is mainly the intangible psychological and emotional benefits that we get from our strong ties. Reliable and trustworthy, bonding capital often protects us from psychological or emotional harm; we seek it out as a means to help lessen our worry or trepidations. When we lack confidence in our abilities or our self-confidence and self-esteem are challenged by new trials, bonding capital obtained from strong ties is capital that we use to regroup and recover our determination and fortitude. For group members of a strong-tie network, bonding capital can have an effect on their solidarity (Williams, 2006), an aspect of bonding capital especially valuable for members of marginalized or outsider groups (Granovetter, 1983). The homophily of an outsider group or a marginalized group in a dominant group's space or institution can provide the bonding capital that is sometimes needed by a member to assuage the psychological and emotional stress experienced. Strong-tie networks can provide bonding capital that serves as a safe harbor for the anxieties often experienced by cultural outsiders and marginalized people. However, it is also true that subjugated groups and group members must also seek out weak ties in order to access bridging capital that is necessary to marshal their forces as a group or to improve their situational context (Williams, 2006).

We establish strong ties with trusted friends and relatives, with people whose social circles overlap with ours and who are often much like us (Gilbert & Karahalios, 2009). Bonding capital is transmitted across the close, trusted relationships of networked strong ties. These networks are often homogeneous (more like ourselves than not) and are often exclusive (Putnam, 2000). The importance of strong ties can be understood through research that shows the inclinations of students to engage, seek support from, and generally stick with members of their own race or ethnic group. This phenomenon has been well documented in K–12 and higher education (Daniel Tatum, 1997; Fleming, 1985). Additionally, it appears that users' ability to segment their networks on Facebook in ways that correspond to weak and strong ties enables them to regulate what they disclose and to whom. In other words, communication across strong ties on Facebook is likely to include more serious or personal information than communication across weak ties (Newman et al., 2011).

Emotional health is positively impacted through strong ties; their intensity and intimacy influence the capital transmitted, and they contribute to our psychological welfare (Gilbert & Karahalios, 2009). Our strong ties require more time and effort to maintain, but because they are so trusted sensitive information is readily circulated through them. Strong ties are fewer in number than weak ties; we can handle more weak ties because they require less intimacy and emotional commitment. In spite of this, weak ties—our acquaintances but not best, closest friends and family—circulate capital that benefits us in our everyday work, our current and future goals, making them truly important to us. As such, weak ties are especially useful contexts for the sharing of campus capital. Golder, Wilkerson, and Huberman (2005) found that the average Facebook user has 144 Facebook friends. According to the researchers, among the 4.2 million North American college users sampled, it appears it is easy to have lots of friends and messaging partners with whom students are likely to have strong ties.

Technology and Social Capital

How can tablet technology and social media serve as the means for FGCS to acquire campus capital? How are these technologies particularly suited for the acquisition of campus capital? How is application software designed to enable the acquisition and circulation of campus capital by FGCS?

Lin (1999) rightly anticipated that "a new form of social capital, cybernetworks" (p. 28), changes the landscape of how, where, and why social capital is transmitted, exchanged, and circulated. More importantly, online social networks could expand opportunities for social capital acquisition and in doing so increase the value of social capital over traditional forms of personal capital. Cybernetworks could empower historically marginalized actors and groups because online resources move quickly across networks that are themselves networked, thus increasing opportunities for resource acquisition and circulation. Lin reasoned that a central feature of cybernetworks is the formation and use of social capital; that "access to free sources of information, data, and other individuals creates social capital at an unprecedented pace and ever-extending networks" that are simultaneously personal and wide-ranging (p. 46). It seems reasonable, then, to consider and examine the ways in which online social networks and mobile technology could be leveraged to facilitate the acquisition and circulation of campus capital by FGCS to advance their individual and group success.

TABLET TECHNOLOGY AS A MEANS TO ACCESS CAMPUS CAPITAL

Tablets are portable computers with a touch interface that allows users to navigate applications quickly and easily. With improved Wi-Fi connectivity, longer battery life, on-board and cloud data storage, and an ever-expanding selection of accessible software, tablet technology enables access to social media and other Web 2.0 applications. Tablets were first conceived as simply portable extensions of desktop computers; now their improved functionality gives users greater and greater access to informational and relational connectivity. Introduced in 2010 as the gold standard of tablets, the Apple iPad allows users to consume and produce videos, text, and images and to interact with application software for information-getting, entertainment, and education. Though favored among tablets, iPad technology is not without its problems. Without a physical keyboard, typing longer documents is arduous; removable, Bluetooth-equipped keyboards can be expensive (typically ranging from $99 to $200 in price). Users often find that friction between applications (a user's ability to move from function to function across applications) is uneven. Without a USB port, extending the functionality and memory of the iPad is limited, and compared to the price of a smartphone iPads are expensive.

Since the introduction of tablets, their use has increased 10-fold. In 2010, only 3% of the U.S. population owned a tablet; by 2015, almost half of the population owned a tablet (Anderson, 2015). The iPad dominates the tablet market. Currently, almost 80 million people in the United States own iPads, but by 2020 that number is projected to rise only slightly (Statista, 2016). The plateauing of tablet devices is correlated with the development and continued enhancement of smartphone technology. Smartphones provide parallel functionality to tablets, are cheaper to purchase, and now support multimedia with more powerful battery life. Smartphones are now owned by 68% of the U.S. population (Anderson, 2015) and their adoption continues to grow at 16% annually (comScore, 2015). Because smartphones have better battery life, cameras, and webpages optimized for smartphone screens, media analysts project that they will replace tablets as the preferred mobile device (How-to-Geek, 2016; Taylor, 2015).

Broadening and improving the range of digital functionality, mobile devices like iPads and smartphones have become preferred technology in the past decade. Internet users have increasingly abandoned desktop technology for mobile devices like smartphones and iPads. Mobile technology use now outpaces desktop computer use with 90% user time spent on apps, and mobile apps are quickly becoming the primary point of Internet access for users (Chaffey, 2015; comScore, 2015). Smartphone and tablet use has increased exponentially since 2011, with an ever-increasing number of mobile-only users, particularly among Millennials (21%)

(Dryer, 2015). Users of mobile technology like iPads and smartphones spend roughly 2.8 hours per day on these devices, dedicating most of their time to social networking apps (Dogtiev, 2015). Mobile applications are now largely consumed on smartphones (comScore, 2015).

Early research suggested that the educational value of a tablet is "its capacity to function as a consolidator or aggregator of information" (Angst & Malinowski, 2010, p. 2). In case studies of iPad use in colleges and universities, researchers found that the iPad appealed to students because it enabled broader information-getting and easier collaboration with peers, provided them with more tools than books, was beneficial for class discussion because it enabled students to navigate quickly to identify passages in texts, and offered overall convenience and flexibility (Angst & Malinowski, 2010; Hall & Smith, 2011; Marmarelli & Ringle, 2011). A Pearson Foundation study (2011) found that 86% of college students who own a tablet believe that it enables more efficient studying and 76% believe that the device has improved their academic performance. Tablet technology appears to provide students with opportunities to access campus information and course content and to communicate with other students, faculty, and staff (Rodriguez, 2011). iPads have been shown to improve and positively affect students' perceptions of the learning experience, and to motivate students to engage in learning (Brand et al., 2011; Kinash, Brand, & Mathew, 2012). However, learning outcomes have not necessarily been positively impacted by iPad use according to early research (Nguyen, Barton, & Nguyen, 2015). There is little evidence to date that iPad use directly and positively impacts learning objectives and/or improves academic performance.

Students report that tablets are simply replacements for conventional learning tools like laptops and pen and paper (van Oostveen, Muirhead, & Goodman, 2011), but it is also the case that tablet technology has been shown to promote their engagement and to broaden communication and learning (Manuguerra & Petocz, 2011; Rodriguez, 2011). Among residential elementary school students, for example, tablets have been used effectively to promote communication with teachers and advisors, a proxy for accessing important social capital in that educational context (Satoh et al., 2015). Among university students in the United Kingdom, tablet technology has positively affected their academic engagement, and more importantly, has increased their autonomous use of the technology to acquire social capital through formal and informal learning conversations (Nortcliffe & Middleton, 2013). As a means to improve low-SES student performance, researchers found that tablets can help decrease the academic effects of low SES among elementary grade students (Ferrer, Belvís, & Pàmies, 2011). This research argues for the value of tablets as an educational "strategy that evidently contributes to the reduction of socioeconomic inequalities" (p. 287). Students' access to information and resources was facilitated by tablet technology, an act of improving the funds and range of social capital restricted and regulated by socioeconomic class.

SOCIAL MEDIA AS A MEANS TO ACCESS CAMPUS CAPITAL

In general, seeking information on online social networks like Facebook is a positive predictor of social capital (Gil de Zúñiga, Jung, & Valenzuela, 2012). Research on social capital acquisition on social media has suggested that these media (in particular, Facebook) are a means to access bonding and bridging capital through network ties (Ellison, Steinfield, & Lampe, 2007, 2011), though the access to bridging and bonding capital is nuanced by user intent and activity level, preexisting relationships, and other situational variables (Brooks et al., 2014; Burke, Kraut, & Marlow, 2011). Facebook, for example, appears to be the social lubricant to facilitate the circulation of useful information (Ellison, Steinfeld, & Lampe, 2011) and where active users can access sources of novel information (Brooks et al., 2014). Though structurally Facebook is a means to access more social capital, it is also the case that how users maintain and manage their network connections determines their perception of what bonding and bridging capital is acquired (Brooks et al., 2014). For example, users who perceive Facebook as a means to obtain information through their networked connections are aware of "the emotional and inclusive bonding capital and the more instrumental and broad bridging capital" (p. 10). Researchers noted that Facebook users show an increase in bridging capital as a consequence of directed (active production) and undirected (passive consumption) communication, but that accrual of bonding capital appears less affected by Facebook use (Burke, Kraut, & Marlow, 2011). It appears that because strong ties are already laden with bonding capital, communicating with strong ties on Facebook to acquire bonding capital is relatively unnecessary, and that using Facebook does not directly increase the value of the connection (Burke, Kraut, & Marlow, 2011).

On social media sites like Facebook, socioeconomic status seems to correlate with access to bridging and bonding capital. It appears that despite the fact that college students from high-SES backgrounds have larger networks and thus more opportunities to access bridging capital, these students appear to better capitalize on their existing networks rather than increasing their opportunities for more and more capital (Brooks et al., 2011). Increasing their opportunities for networking beyond their large and dense networks in order to increase their capital does not appear to correlate with high-SES student users. Yet, for FGCS students who typically come from low-SES backgrounds, it appears that social media do enable them to purposefully extend their networks in order to access bridging capital about college-going, for example (Wohn et al., 2013). Moreover, for FGCS and students of color (SOC), social media may be a means to access support—bridging and bonding capital—that positively affects their social adjustment in college (Gray et al., 2013). By using social media, college students appear to gain positive outcomes in developing their identities and integrating into the campus (Junco, 2014), no doubt a function of accessing campus capital through these media.

This chapter presented the conceptual framework and rationale for an intervention designed to better the college experience of FGCS and to improve their chances for college success. Campus capital is offered as a way to understand the information and knowledge or social capital that is necessary to successfully navigate campus norms. Grounded in the principles of critical theory and social capital acquisition, campus capital conceptually organizes information, know-how, expertise, awareness, sensibilities, and competencies—the social capital of college-going—that can be acquired and circulated by FGCS to advance and better their college years.

2

Being First-Gen on Campus

For as long as he can remember, Nico has heard from family members and friends that he was going to be the first person in his family to go to college. Every quarter, when he brought home a report card, or received an award, his father would express his pride in his son and describe how his family had sacrificed greatly to emigrate from Honduras to America. It wasn't until his high school counselor helped him with his legal status, which had been pending for a long time, that he actually started to believe it was possible. Still, even with his new green card and an admissions letter from Millbrook University, he wasn't fully convinced that he would ever accomplish the goal of being a college graduate. He hoped that feeling would go away upon enrolling, but even at the end of his first year, Nico wondered whether he would ever fulfill his father's dream. Each time he encountered a hardship, this uncertainty loomed. At first, it was financial stress. Although Nico earned a merit scholarship, there were many bills to pay, prompting him to work about 15 hours a week in his father's auto body shop. As valedictorian of his high school class, Nico had come to see himself as academically capable. However, he found the academic demands at Millbrook much harder. Sometimes he became lost in lectures when faculty referenced ideas or concepts Nico had never heard of. However, his greatest frustration came when he believed he understood the course content only to earn a low grade on the final paper or exam. Yet, with each challenge, Nico worked harder.

At the start of his sophomore year, Nico still feels the weight of each of these worries. Some days, he feels more sure of himself. He's even caught himself saying to himself, "I got this." Yet, it seems that no matter how many obstacles he overcomes, he continues to encounter new, unexpected ones. For example, the other day, Nico overheard two classmates talking about summer internships in Washington, DC, for public policy. Although Nico was a Finance major, he wondered whether he was supposed to be doing an internship and if so, were any available in finance? And, were they paid or did students have to pay? Nico knew he couldn't afford to not work in the summer; however, he also didn't want to miss out on something that would be helpful to his future. He had recently joined a men's mentoring group and figured he might ask some of the seniors in the group what they knew.

Like Nico, many first-generation college students (FGCS) manage the sometimes bumpy road to higher education only to find that once there more challenges exist. The academic, social, financial, and personal challenges faced by FGCS, which often result in lower degree completion, have put a spotlight on this unique population of students. National interest in increasing the number of Americans who attend and graduate from higher education has brought about a range of new support programs, improved federal policies, increased federal aid, and other investments geared toward ensuring that the pathways to college are clear for all students, especially those previously underrepresented on college campuses. For FGCS this has led to improvements in the pre-college experiences that set students up for postsecondary success. Simultaneously, institutions have been charged with establishing campus programs and supports that promote degree attainment among a population of students who, research suggests, are vulnerable to disengagement or dropping out. However, as we see in Nico's situation above, FGCS experiences in college are fraught with issues that span financial, academic, social, familial, and personal domains. Ensuring that students like Nico can successfully navigate these challenges will require that institutions design programs and policies that take into account FGCS unique experiences and needs.

More than two decades ago, London (1989) called on institutions of higher education to possess a "keener understanding of the sensibilities and concerns [FGCS] bring with them and of the difficulties they encounter" (p. 168). The following decades of research brought greater understanding of FGCS experiences on campus and subsequently new programs and intentions of support. As we detail in this chapter, there is a considerable body of literature describing FGCS students and how their transitions to college differ from those of continuing generation college students (CGCS). As a result of that scholarship, institutions have engaged in a range of practices and programs to foster campus engagement and academic attainment among FGCS. Yet, despite these efforts, degree attainment rates of FGCS continue to lag behind those of their peers (Chen, 2005; Engle & Tinto, 2008). Moreover, even on college campuses where FGCS like Nico are successful academically, their social and personal experiences are compromised in varying ways (Davis, 2010).

In this chapter, we describe FGCS and their unique experiences in higher education. We begin by providing a clear definition of FGCS in the context of our research/study, clarifying a term that is used widely for a diverse group of students. Next, we portray the landscape of college enrollment and degree attainment among FGCS, specifically calling out the disparity in a range of student outcomes. Then, we describe how FGCS academic, financial, social, and personal experiences on campus necessitate the type of web technology intervention we present in this book. Finally, we share current efforts on college campuses to support and promote FGCS success in higher education.

Who are FGCS?

The term "first-generation college students" is widely used today despite many differences within this group. From the former First Lady Michelle Obama to students like Nico, there is great diversity among this widely used descriptor. Most commonly, this term is used to refer to students who are the first in their family to attend college. However, even this definition carries some variability. For example, some researchers and programs differentiate among FGCS, with criteria inclusive of students whose parents or siblings have never attended *any* form of postsecondary education (Chen, 2005). In other programs, such as Federal TRIO Programs, first-generation students are identified as those for whom neither parent has earned a four-year degree (Engle & Tinto, 2008). In our study we are defining FGCS as those students with a parent who has not earned a bachelor's degree, similar to the definition of Federal TRIO Programs, However, beyond a family member's experience in higher education, there is a great deal of diversity among FGCS.

There is a range of characteristics that are highly correlated with being first-generation, but most certainly not synonymous with it. FGCS are typically more likely than their CGCS peers to be from ethnic minority groups or from low-income communities (Engle & Tinto, 2008). In their analysis of low-income FGCS populations, Engle and Tinto found that more than half of these students are racial/ethnic minority students. FGCS are also more likely than students whose parents attended college to be from low-income families (Chen, 2005; Warburton, Bugarin, & Nuñez, 2001). For example, Chen found that half of FGCS are in the lowest-income quartile, compared to less than 10% of peers who are continuing generation students.

Another characteristic that distinguishes FGCS from their continuing generation college-going peers is their precollege experiences. Similar to the precollege experiences among low-income students, many FGCS attended schools that lacked strong college-going cultures and a rigorous college preparatory curriculum (Jehangir, 2010a). Specifically, because FGCS often hail from low-income communities, unlike their upper-income peers, they may not have had access to high-quality college counseling, which helps students find a good fit and college match. In addition, these students are less likely to have planned for college early or to have adequate access to comprehensive information, and are more likely to have limited support in the college choice process (Institute for Higher Education Policy, 2011). These precollege experiences shape FGCS transition to college and ultimately their experiences there.

Despite the overlap in income, race, and precollege characteristics among FGCS, there is great variation among this group, which makes writing about them as a singular group challenging. As a result, many researchers characterize a specific subgroup of FGCS as their point of focus. For example, Jehangir (2010a) wrote about low-income first-generation (LIFG) students to capture how first-generation and low-income status creates marginalization and isolation in a learning community. Social class, which is linked to income, also has relevance to their transition to college. Similarly, McCorkle (2012) has examined the experiences of first-generation African American students, considering the ways in which race and first-generation status shape students' perceptions of and experiences in college. In this book, we use the broader "FGCS" term to describe students who were part of our project, which primarily included students whose parents have not completed a bachelor's degree and who are primarily low-income students of color (SOC).

FGCS ACCESS TO AND SUCCESS IN HIGHER EDUCATION

At a time when the possession of a college degree is critical for entry into the middle class, as well as the economic security of the United States, uneven degree attainment among subgroups of students is of great concern. First-generation college students constitute a large portion of students who have fallen out of the K–16 pipeline, thus creating a swell of interest in this population. The transition to and through college is precarious for FGCS with documented gaps in enrollment and degree attainment between this group of students and their continuing generation college-going peers. Such differences in performance are evident across a range of educational benchmarks such as GPA, enrollment in developmental education, persistence rates, and even satisfaction in college (Chen, 2005; Engle & Tinto, 2008). Together with the implications of stopping out of college for students, statistics paint a concerning picture.

Locating specific data on this population is difficult due to varying definitions used by researchers, and related research focusing on similar, but not synonymous identifiers such as race or income. Nonetheless, like the students who participated in our project, this group of students represents a large one on college campuses today. Among undergraduates enrolled in college in 2012, one-third of them came from families where at least one parent held a high school diploma, and another 28% had at least one parent who had only some postsecondary experience and did not complete a bachelor's degree. As previously mentioned, many racial and ethnic minority students are also the first in their family to attend college. In fact, FGCS make up 48% of Hispanic students, 42% of Black students, and 40% of Native American students (U.S. Department of Education, 2013). Low-income FGCS, the focus of our book, make up close to one-quarter of the undergraduate population, and are often from racial and ethnic minority groups with lower levels of academic preparation (Engle & Tinto, 2008; Tym et al., 2004).

Although the past decade has seen an increase in the number of FGCS enrolling in college, degree attainment rates among this group continue to lag behind those of their peers whose parents attended college. For starters, FGCS are less likely to graduate from college and more likely to leave in the first year than their peers (Chen, 2005; Engle & Tinto, 2008). Specifically, 26% of low-income FGCS left college during the first year compared to 7% of continuing generation students who were not low-income (Engle & Tinto, 2008). In Chen's study (2005) of high school graduates, only 24% of FGCS graduated with a bachelor's degree within eight years of high school graduation compared to 68% of continuing generation students. Among low-income FGCS, that number falls even lower, with only 11% earning a bachelor's degree. One exception to these gaps is at elite four-year institutions where the graduation rate for FGCS and their continuing generation peers is at least 85% (Bowen, Kurzweil, & Tobin, 2005). Although these students may be graduating from these elite institutions at greater rates, their social and personal experiences are not necessarily positive, as we will examine in the following section.

FGCS Experiences on Campus

> Even though I know sometimes that I can't help it that I have to be there to help my family, try to help them and support them financially at times, it's like I wish I could be so much more present on campus, especially with me being able to live here for the next school year for free, I want to be able to fully utilize that. But it's like at the same time I have to understand that I have to be present with my family too, so it's kind of like what's more important.
>
> —Mason

Historically, academic underpreparedness was believed to be the primary culprit behind the gaps in degree attainment among FGCS. However, the scholarship on this topic is far more varied with greater attention to the specific needs and experiences, including the precollege experiences, of this population. Much research has been dedicated to the challenges faced by first-generation college students. Here, we discuss FGCS unique academic, financial, social, and personal experiences, and importantly, how these understandings warrant particular institutional efforts to promote academic success and a positive college experience.

ACADEMIC EXPERIENCES

> We all come from a lower high school, not as good as the people, the regular students here so we struggle a lot with just adapting to the academic here, to the classes. Most people took five AP classes or three at least. I personally didn't take any.
>
> —Nico

As Nico describes, the academic expectations of college can leave students feeling as though their previous schooling experiences failed to prepare them for the rigors of college. Indeed, this is the most commonly cited concern for FGCS. Mentioned previously, many FGCS come from low-income communities and attended under-resourced and underperforming schools. The statistics on

FGCS academic performance in college confirm the long-term implications of inadequate schooling.

Beginning at the entrance to college, FGCS at 4-year colleges score 122 points lower (1054) on the SAT than continuing generation students (1176). Indeed, these students are at a disadvantage compared to their continuing generation peers in terms of admission to selective colleges (Saenz et al., 2007). Interestingly, attending a four-year college appears to mitigate the degree to which lack of preparation hinders long-term success. FGCS who start at four-year institutions are seven times more likely to earn a bachelor's degree than their peers starting at a two-year public institution (Saenz et al., 2007).

Gaps in academic preparation for college are also evident in AP course-taking with FGCS typically less likely to have taken AP courses and tests. This is partially attributable to attendance at under-resourced schools that historically fail to offer AP courses. However, among FGCS students who did take AP exams, they tend to have lower scores on average than their continuing generation peers (Balemian & Feng, 2013). Another important indicator of academic underpreparedness is enrollment in remedial or developmental coursework. These courses are designed to compensate for inadequate preparation and act as a bridge to the academic demands of higher education. More than half of FGCS enroll in remedial coursework (Chen, 2005; Engle & Tinto, 2008) in higher education. Enrollment in these courses has been of great concern due to the need to use financial resources for these non-credit-bearing courses. Continued academic performance gaps among FGCS emerge during students' first year, noticeably in lower average GPAs compared to those of their continuing generation peers (Chen, 2005; Vuong, Brown-Welty, & Tracz, 2010). Relatedly, FGCS earned 18 college credits per year on average compared to continuing generation students, who earned 25 credits on average (Chen, 2005).

These disparities, such as reduced credit accumulation and low performance, have been attributed to disrupted enrollment and part-time attendance (Chen, 2005), low levels of self-efficacy (Vuong, Brown-Welty, & Tracz, 2010), limited involvement with faculty (Kim & Sax, 2009), and inadequate preparation in high school (Engle & Tinto, 2008). Because FGCS typically work more hours than their peers, and are less likely to live on campus (Pascarella et al., 2004), their opportunities for seeking academic support are often compromised. Moreover, according to Kim and Sax (2009), FGCS are less likely to be involved in faculty research, to communicate with faculty outside class, and to engage with faculty during lecture class sessions compared to their continuing generation peers. In light of research that has confirmed the benefits to persistence that occur when students are engaged in the college experience, it appears that FGCS are clearly at a disadvantage because of their limited engagement and connection to faculty.

FINANCIAL CONSIDERATIONS

> [I'm] working to pay my tuition. It's not necessarily that much money. It's just the work that I do is in the dining hall and athletics, so it's just like having two jobs. In the beginning of the year I didn't want to work in dining, but I had to because athletics they weren't giving me enough money. . . . It's kind of hard to keep working when I have exams. I had to switch my shift a lot of times and my boss gets mad. Then they ask me to pick up more shifts, but I really can't because I'm trying to improve my grades. I really don't want to work there anymore, but I have to because that's how I get by.
>
> —Gabriel

As we noted earlier, many FGCS are also low-income, which explains the range of financial issues that influence their college experience. On average, low-income, first-generation students rarely receive the aid necessary to fill the gap between college costs and their estimated family contribution (Advisory Committee for Student Financial Assistance, 2001; Engle & Tinto, 2008). The financial burden on these students and their families that comes with tuition and related higher education expenses can represent half of their median annual income. Making matters worse, the majority of aid received by students comes in the form of loans, and as a result, low-income FGCS students are left with higher loan indebtedness than other students (Engle & Tinto, 2008).

As Gabriel's quote reminds us, the financial issues FGCS face are certainly not limited to their tuition. Financial obligations and additional college expenses require that FGCS work while in school. Almost two-thirds of FCGS work more than 20 hours a week compared to less than half of continuing generation students who are not low-income (Engle & Tinto, 2008). In addition to working many hours, these students often find themselves unprepared for the hidden financial costs associated with being social and engaged on campus. In their recent book, *Paying for the Party: How College Maintains Inequality*, Armstrong and Hamilton (2013) described the ways in which low-income students miss out on key opportunities to participate and socialize on campus due to the costs of partying cultures. For students in our project like Gabriel, the need to work and limited financial resources preclude opportunities to participate in academic and social experiences in college. Given the context of privilege within higher education, it is understandable that the socioeconomic dimension of identity often becomes especially significant for FGCS as they transition and adapt to college life (Kezar, Walpole, & Perna, 2014).

CAMPUS CAPITAL/KNOWLEDGE

> I didn't . . . like, I didn't know what office hours were, so I didn't really use that last semester at all and I should've. But this semester, I've gotten pretty good so now [faculty] know me pretty well because I go there a lot.
>
> —Krista

Beyond academic and financial factors, FGCS also enroll in higher education lacking essential college knowledge that is otherwise possessed by their peers whose parents attended college. For Krista, this knowledge included understanding the specific purpose and format for office hours, something Cox (2011) described in her book, *The College Fear Factor*. Whether it is how to utilize office hours or how to operate bureaucratic processes, continuing generation college students have the benefit of parental knowledge and experience regarding higher education that gets transmitted across generations. On the other hand, FGCS have not been recipients of the kinds of transmitted knowledge that comes from the previous experiences of family and close family friends. The vast majority of programs designed for FGCS focus on how to access college, with little attention paid to the type of guidance and information needed once enrolled (Institute for Higher Education Policy, 2012). As a result, many students in our project describe a lack of familiarity with specific institutional offices or the general working systems of higher education.

Case studies and accounts of FGCS typically reveal students' lack of familiarity with how things work, as seen above in the quote by Krista. In their book, *After Admission: From College Access to College Success*, Rosenbaum, Deil-Amen, and Person (2006) referred to the "social prerequisites" that are necessary for success in community colleges. They suggested that colleges "demand a certain level of social know-how, a set of skills and knowledge that help students understand school procedures and navigate these institutions" (p. 1130). Fortunately for Krista, these expectations and processes became clear and she was able to benefit from these intended resources. The same realization came from students profiled in Cushman's (2006) *First in the Family* series. In the students' words, their initial experience eventually gave way toward a new understanding of codes and norms. However, specific knowledge and skills are not always readily apparent and can leave FGCS feeling as though they do not belong.

Specific terminology to describe this type of knowledge or skills has been introduced in different ways. In the context of precollege knowledge, Bell, Rowan-Kenyon, and Perna (2009) used the term "college knowledge" to include college cost and aid, academic requirements, and the education needed to meet career aspirations. Rosenbaum, Deil-Amen, and Person (2006) referred to these as "social prerequisites." Each of these terms draws on notions of cultural capital to describe the knowledge and behaviors used to access opportunities and information. This type of capital requires specific types of relationships, often referred to as social capital, and thus many researchers utilize Bourdieu (1986), Coleman (1988), and Lamont and Lareau (1988) to illustrate the kinds of capital associated with the transitions to college and how institutions transmit it.

SOCIAL AND PERSONAL EXPERIENCES

> I've been homesick, trying to find my niche
> of people here to hang out with.
> —Ava

A critical aspect of FGCS experience includes their social and personal development and the ways in which their emerging identities shape their interactions with peers and the larger institution. The push and pull felt by FGCS, like Mason described in the quote earlier in the chapter, can make it difficult for students to feel at home in college while also remaining connected to one's home community and culture. In many ways, FGCS are actively straddling two critical sociocultural contexts that inform identity development and their broader college experience. They struggle to become a part of a college environment with a focus on formal education and new norms and cultures, while remaining connected to home/community contexts that are not privileged in terms of educational status (Orbe, 2004). According to Ava, it can be hard to find one's place on campus as a FCGS. Because these social relationships inform students' identity and personal development, students' experiences with others and the meaning they make of them are central to FGCS experiences in higher education.

Dramatic changes to students' sense of self in college are common among all students. However, this process can be complicated for FGCS, who are negotiating multiple ascribed dimensions of their identity such as race and class, as well as beliefs in ability due to inadequate academic preparation. Aries and Seider (2007) found that social class shapes FGCS identity and thus their ability to feel a sense of belonging and adjustment to college. These authors argued that when FGCS move into contexts of privilege, such as universities, this transition "has significant effects on one's sense of self because identities must be renegotiated" (p. 139). According to Olson (2011), FGCS experience a higher level of social class dissonance compared to non-first-generation students. In some cases, FGCS may actively hide their class background in an effort to "fit in" with the middle class (Martínez Alemán, 1999, 2000). This sense of self and broader process of identity development has implications for how students form social relationships on campus. For example, research shows that dissonance may make it difficult to build new networks, leaving students to rely on old networks and resources (Saunders & Serna, 2004). Recognizing that students' new peers in college possess types of cultural capital that may be useful to them, failure to form these relationships puts FGCS in a risky position.

Perhaps because of their identity negotiations and questions about belonging, FGCS show lower levels of participation in traditional academic and social experiences. Many studies have found that this group of students is less likely to participate in traditional and normative engagement measures such as faculty interaction and co-curricular activities that are known to promote persistence in college (Lohfink & Paulsen, 2005; Nuñez & Cuccaro-Alamin, 1998; Pascarella et al., 2004). In some cases, these lower levels of engagement can be attributed to logistical features of their experience. For starters, FGCS are more likely to live at home. These students are more likely to live within 50 miles of home (35%) compared to CGCS, and there is also a 14 percentage point difference in choosing to live on campus, with FGCS (69.3%) less likely to live on campus at four-year institutions than their continuing generation peers (84.5%) (Saenz et al., 2007). According to St. John, Hu, and Fisher (2010), traditional forms of engagement are often a challenge for FGCS who may need to focus more on academics to make up for their lack of academic preparation. That these logistical barriers stand in the way of engaging on campus is especially troubling in light of studies that link reduced participation rates to lower persistence in college.

Although these logistical challenges are sufficiently obstructive to campus engagement and social networking, research suggests that much more is likely to be shaping the ways in which FGCS relate to their classmates and engage in campus programs and activities. Jehangir (2010a) attributed FGCS lack of involvement and engagement to feelings of marginalization and isolation on campus. In her book, *Higher Education and First-Generation Students*, she noted that some FGCS experience marginalization before attending college as they "see themselves as outsiders in the educational context even before they arrive to college" (p. 20).

Cultural dissonance also explains why some FGCS might find themselves struggling to connect socially or academically on campus. Stephens et al. (2012) described the challenges students face in navigating the college experience using the cultural mismatch theory. They hypothesized that this mismatch is the root of the achievement gap for first-generation students. Since institutions are primarily focused on norms of independence, where students realize their own potential by separating from family, the authors theorize that first-generation students do not perform at the same level as continuing generation students since FGCS are more familiar with interdependent norms they brought from working-class families, focused on building community and adjusting to the needs of others. In testing this theory, they found that this achievement gap can be minimized if institutions can reframe university culture to include working-class norms. Vasquez-Salgado, Greenfield, and Burgos-Cienfuegos (2014) found this to be especially true for Latino/a FGCS who were raised to make family obligations a priority. Conflicting school and family demands affected student well-being and produced family tension and turmoil within the students. FGCS, and Latino/a FGCS in particular, had a higher likelihood of family achievement guilt than their peers (Covarrubias & Fryberg, 2015).

With some variation, the constellation of factors that shape FGCS experiences in the preceding domains (academic, financial, informational, and social/personal) take place on all types of college campuses. However, our project took place at Millbrook

University, a pseudonym for a private, highly selective, four-year, Predominately White Institution (PWI). It is worthwhile to understand the unique experiences of FGCS students at PWIs, termed as such to describe institutions in which White students account for 50% or more of the entire student enrollment. Certainly the experiences described, especially academic readiness, financial burdens, marginalization, and isolation, to name a few, may be heightened for FGCS attending a PWI (Brown & Dancy, 2010). For example, we have already described FGCS as often coming from ethnic and racial minority groups, where evidence suggests they will experience various forms of racism and isolation (Quaye, Griffin, & Museus, 2015). In addition, the class dissonance felt by FGCS is pronounced at selective, private colleges where cultural values and expectations may make it difficult for FGCS to connect socially with other students and faculty. However, our interest in uncovering responsible practices and policies that alter FGCS experiences at PWIs extends beyond these heightened experiences. First, because degree attainment and educational performance are higher at institutions with greater selectivity, they are important places for FGCS to attend and succeed (Melguizo, 2008). Second, recent national headlines calling attention to the propensity of low-income students to undermatch or fail to apply to the most selective school for which they are qualified, and accompanying interventions designed to help these students match at the most selective possible suggest that it is even more essential that we prepare practitioners at PWIs to support FGCS students. Third, these selective institutions act as pathways to prosperity due to the social networks of alumni and the pipeline to graduate work opportunities they possess. There is no question that all postsecondary institutions would benefit from a greater understanding of strategies that show promise in addressing the unique experiences of FGCS. However, our project examined the potential for using technology to enhance institutional supports at a PWI, knowing the possible broad application it holds for promoting engagement and belonging for FGCS in higher education.

What Institutional Programs and Practices Are in Place to Meet the Needs of FGCS?

We have seen a surge of new programs and practices across various institution types designed to improve the persistence and graduation rates of FGCS (Davis, 2010). Without question, vast improvements have been made to the precollege experiences of FGCS in the form of many college access and preparation programs (Savitz-Romer & Bouffard, 2012). These programs seek to improve students' academic preparation, expand their information and knowledge of college, and increase access to financial aid and scholarships. There is also a growing number of programs that are also including a focus on the social and emotional skills of students (Savitz-Romer, Rowan-Kenyon, & Fancsali, 2015). At the higher education level, a concerted effort is under way as well. First, there is a set of programs designed to support the logistical challenges faced by first-generation college students. In addition, institutions host various support programs that create opportunities for building social capital-rich relationships among college SOC (many of whom are FGCS) to aid students in accessing social networks and cultivating social capital through a network of diverse and integrated connections early in their college experience, which is of critical importance (Museus, 2010).

Perhaps most well-known among these types of programs are longstanding Federal TRIO Programs, which are designed to provide a range of academic and social supports for FGCS, as well as other students who meet criteria believed to put them at risk of disengagement or dropping out. Although Federal TRIO Programs have been designed to support this group of students, these programs have the capacity to reach only a small number of students who may be eligible (Kezar, Walpole, & Perna, 2014).

Institutions have launched structural programs designed to augment FGCS academic experiences, such as learning communities and peer-mentoring programs. Interest in these types of initiatives emerged in response to a wide body of research regarding the barriers facing FGCS, which has been grounded in traditional conventions of engagement and the ways in which students claim membership or belonging in the campus community. As previously mentioned, FGCS experience lower documented levels of engagement than students whose parents have completed college (Kuh et al., 2007), and these programs are designed to build relationships with faculty and peers and to increase rates of academic success on campus.

Higher education institutions have also utilized peer learning programs to foster the kinds of social relationships that lead to increased engagement and the provision of social support, not typically sought out by FGCS. Dennis, Phinney, and Chuateco (2005) found in their longitudinal study of ethnic minority FGCS that peer support is a stronger predictor of success in college than family support. Students relied on their peers for instrumental support, which was not often available from family members who have not attended college. The transmission of these norms and college-staying behaviors through peer relationships takes on the form of academic capital that can serve to promote students' engagement and success (Savitz-Romer & Bouffard, 2012).

Policies and practices found to promote FGCS success in college also include faculty-targeted initiatives. This strategy has been popular in light of the fact that faculty members are often students' first point of academic contact. Initiatives in this vein include faculty serving as case managers and advisors of support programs, faculty professional development, and changes in faculty responsibilities to include teaching nonacademic courses. Curricular and pedagogical reforms are also important to create an engaging environment in the classroom to promote student success (Institute for Higher Education Policy, 2012; Kezar, Walpole, & Perna, 2014). Recently, researchers have started exploring how bringing the lived experiences of FGCS into the classroom, and other interventions, can boost student self-efficacy and academic performance, and attempt to close the social class achievement gap. These experiences enable FGCS to be co-constructors in knowledge production, and help to sustain students who have previously been marginalized in the classroom due to their first-generation status (Jehangir, 2010b). Harackiewicz et al. (2013), in their attempt to combat stereotype threat for FGCS who were majoring in science, technology, engineering, and math (STEM), used a values affirmation intervention in a double-blind randomized trial experiment with more than 700 students in an introductory biology sequence (approximately 20% FGCS and 91% White) and found that the values affirmation intervention narrowed the social class achievement gap for FGCS. Students who were part of the intervention were more likely to do well in the class and other grades in general, and to stay on track for other science-sequence courses. In a randomized controlled trial using a difference education intervention to close the social class achievement gap for FGCS, Stephens, Hamedani, and Destin (2014) found that those students who were in a section where panelists highlighted first-generation status had higher GPAs than other first-generation students, and shared that the panels helped students to see FGCS juniors and seniors who were successful at the institution. This intervention reduced the social class achievement gap by 63%.

One program at the institution at which our project is based is designed to decrease feelings of marginalization and increase a sense of community on the campus for this population of low-income students, who are primarily FGCS. Students are provided opportunities to attend campus events (e.g., sporting events, concerts, service trips, and retreats) for free through a campus collaboration designed to enhance the experience for these students who would not be able to attend these co-curricular events, which are a large part of the campus culture, without financial support. This opportunity increases the ability of students to benefit from campus events that other students take for granted.

This chapter has provided an overview of FGCS and their experiences in college. These students are pioneers within their own families in their quest to earn a college degree. They face multiple challenges in terms of precollege preparation and resources, differential engagement and experiences while in college, and the lack of valuable campus capital that their continuing generation peers have, which decreases their chances for success and degree attainment.

3

Web 2.0 Technologies on Campus

WITH CONTRIBUTIONS BY ADAM GISMONDI, KEVIN GIN, SARAH KNIGHT, JONATHAN LEWIS, & SCOTT RADIMER

As Arun enters the hall where his sociology class is taught, he sends one final text to his sister. He knows he needs more time to chat with her about her struggles at home, but he wants to be on time for class. He slips into class, finding the same seat he had last week and reaches into his bag for his iPad and portable keyboard. Both were given to him by the Educational Opportunity Program (EOP) for use at Millbrook, which Arun really appreciates since he does not have a laptop. The professor is at the lectern surveying the room and glancing at her lecture notes, but from the noise level it seems like she will not begin for another couple minutes. Arun does not really know the other students near him, so he decides to check Facebook. The professor raises her voice to signal that everyone should take a seat, and then reminds the group that, per course policy, laptops, tablets, and phones must be put away during class. She looks directly at Arun and a few other students who have devices on their desks while making this announcement.

"This is annoying," Arun thinks to himself, as he sheepishly stows his iPad and keyboard back in his bag and digs around for some paper and a pen. Arun is much faster at typing than handwriting—his writing is practically illegible anyway—and no one seemed to care if he used his iPad to take notes during his EOP courses. In fact, he remembered, for one course an iPad or laptop was necessary, as the instructor regularly referenced assigned e-books and specific apps related to the course content during class. She even suggested meeting with Arun once over Skype when he had a question about a paper the day before it was due. He was surprised by the offer and went ahead with it because it was her idea and he had already missed her office hours for the week. The virtual meeting turned out to be helpful. Refocusing on sociology, Arun makes a half-hearted attempt to write down a few things the professor says, but mostly he just listens or doodles on his paper.

At the end of class, Arun snapped a picture of the whiteboard in an attempt to remember some of what had been discussed and then left quickly to avoid any potential interaction with the professor. Later that day, someone who sat near him in sociology sent him a message, asking Arun to text her one of the pictures he took of the whiteboard. He thought for a few seconds what might happen if his professor found out—is this violating her policy?—but then dismissed the thought and sent her the picture. He does not really know this student and figures it might be a good way to start a friendship.

Like his peers, Arun is a first-generation college student (FGCS) whose life is mediated by ubiquitous technology. He texts his sister on his cellphone, "talks" to his mom on FaceTime, and follows his high school friends and favorite musicians on Instagram. On campus, he uses his tablet in class when professors allow it, to watch movies on Netflix or shop on Amazon in his residence hall room, and to play games in his downtime. Determining college student use of social media and other web-based technologies is an ever-changing proposition, as both the media and the demographics of higher education change continuously. Yet, are we leveraging technology to support the transition to college among a generation that is so deeply technologically engaged?

As we established earlier, knowing how to access resources, how to gain the information or knowledge to facilitate the college experience, is critical for all students but especially for FGCS. Campus and college knowledge and resources can be understood as social capital, or "campus capital" as we have termed it, and their circulation on college campuses happens through networks of relationships outside and inside the classroom. As researchers have identified, users of social media receive bridging and bonding social capital through their network ties (Ellison, Steinfield, & Lampe, 2007; Ellison, Lampe, & Steinfield, 2009; Steinfield, Ellison, & Lampe, 2008). For FGCS, these networks extend to both extracurricular and academic life on campus. Given the ubiquity of these Web 2.0 technologies and students' generational inclination to use them, their impact and effect on the acquisition of campus capital seems likely and positive.

Web 2.0 technologies serve to develop opportunities for forging weak, or bridging, ties outside the classroom. Weak ties are generally newer social connections that occur between people who bring different resources and information to the interaction (Granovetter, 1973). Within the college environment, these connections can occur through programs that set up purposeful interactions, like an orientation program or campus housing activity. As weak ties are often between people with differing backgrounds and personal resources, the ties among these individuals can serve as a bridge to new information and resources that would be challenging to acquire otherwise. The challenges presented by higher education environments to FGCS may be tempered by the acquisition of these new resources.

Social media can play a role in the attainment of social capital among college students, and this is a process that scholars are now beginning to understand (Ellison, Steinfield, & Lampe, 2011; Valenzuela, Park, & Kee, 2009). The personal network of a college student can be largely reshaped through the use of social media, as these new tools can help lower or remove cost barriers to communication (Ellison, Steinfield, & Lampe, 2011). Recent scholarship indicates that social media applications and websites, like Facebook, promote the development of ties to sources of social capital through the ease of relationship building online (Ellison, Steinfield, & Lampe, 2011). The various pathways to online communication on social media result in new relationships for users that then translate into social capital formation.

The social capital that is formed from the connections between individuals who have dissimilar characteristics are of major importance for higher education, as it aids the promotion of diversity as a value and also can serve as a point of exposure to cognitive dissonance, which promotes development (Putnam, 2000; Saguaro Seminar, 2000). As students are exposed to peers of differing backgrounds, from which new experiences, beliefs, and opinions are brought to an interaction, personal thoughts are often challenged. The cognitive complexity that these interactions require can be developmentally impactful.

Co-curricular activities provide universities with the opportunity to engage students in situations that promote such social capital. Traditional activities including interfaith dialogues, diverse speakers, cultural celebrations, sociocultural debate forums, and hall mixers represent the types of organized events that can spur the circulation of capital within a college community. The rise of social media adds a new layer to the developmental opportunities for higher education leaders. Traditional events targeted at circulating and transmitting social capital can now be made available online; forums can be presented and moderated within web communities; and information can be created, presented, and shared easily and without cost through these new media.

Correspondingly, Web 2.0 technologies can also impact FGCS acquisition of capital inside the college classroom. When we consider the effects of social media as instructional technology in the college classroom, it stands to reason that students would have access to greater funds of information and ideas, and grow greater funds of capital through these media. In the college classroom, social media provide students with opportunities to build networks with other students (often different from themselves) through which they can share information and exchange ideas. These social ties between students are made through the communication conducted on social media, communication that ultimately widens and extends individual students' academic and intellectual community and power. Like all social capital, academic ideas and knowledge are transmitted and circulated. Through social media, students can strengthen their relationships with other students and, as a consequence, access greater funds of capital (Ellison, Steinfield, & Lampe, 2007). It seems that social media as instructional technology can help develop relationships necessary for FGCS college success.

In this chapter, we explore the current landscape of Web 2.0 technologies on the college campus, including use among students as well as within the institution. First we present the specific geography of Web 2.0 for FGCS. Then we explore the characteristics of these media inside and outside the college classroom with a specific focus on the capacity for these technologies to positively influence the student experience.

The Landscape of Web 2.0 Technologies on Campus

The emergence of "Web 2.0," a term coined by O'Reilly (2005), reflected a maturation of the Internet in the late 1990s and early 2000s. During this period, the web became a place that encouraged sharing of information, and new software worked to add a new dynamism to the user experience. The earliest forms of the Internet presented a user experience that relied on user-to-website relationships. As the medium developed, Web 2.0 brought about an era of content sharing among users, and this shift is perhaps best characterized through the features of social media that exist today. Both consumptive and productive, Web 2.0 is the broad set of ideas and software characteristics that enable and encourage sharing, participation, and connections among users, while social media are the particular websites and applications that leverage this philosophy within specific outlets. Facebook, Instagram, YouTube, and Wikipedia would all be considered forms of social media that exemplify core principles of the Web 2.0 movement.

It is the development of Web 2.0 and its affiliated forms of social media that have shaped the most recent decade of the user experience online. These developments within technology, along with the devices that allow users to access these media, like computers, tablets, and smartphones, have also had a profound impact on higher education. Students, faculty, and staff within universities have had to navigate the opportunities and challenges that come with these new technologies both in and out of the classroom. Higher education leaders work to understand technology use patterns in Web 2.0 landscapes that grow exponentially and at a rapid pace. Higher education administrators, staff, and faculty seek to successfully implement these tools in a way that facilitates learning and development inside and outside the classroom.

WEB 2.0 TECHNOLOGY USE BY STUDENTS

> I woke up, opened the iPad, I check the [school] app for the weather to see the weather, and then I checked my email to see if there were any emails about class being cancelled or anything, and I starred the ones that were important . . . Then I went on, I think, Instagram . . . put pictures on Instagram from the iPad. . . . I had a discussion so I recorded some stuff from the discussion because I missed something. And then I emailed my professor to ask her if I could do an extra credit assignment, and she emailed me back and was like yes. And, I listened to Pandora.
>
> —Gabriel

College students like Gabriel rely on technology for a variety of purposes, but with a constancy that marks this generation. U.S. college students, especially those on traditional residential campuses, represent a principal share of the Web 2.0 user population. Recent data from the Pew Research Internet Project reveal that young adults ages 18–29 constitute a significant share of the mobile and smartphone user market, as well as a critical share of users of the leading social networking sites Facebook, Instagram and Twitter (Lenhart, 2015). Among young adults, smartphones are pervasive—83% own one. Many of these young adult populations are heavily dependent on their smartphones for Internet access because they are less likely to own another computing device. Approximately 15% of adults ages 18–29 are heavily dependent on their smartphones for online access; 12% of African American and 13% of Latino/as report that their smartphones are their only means to access the Internet. Low-income users are also often heavily dependent on their smartphones for Internet access (Pew Research Center, 2015). As decidedly mobile users, young adults also look to tablet technology for their mobile connectivity. Roughly 48% of young adults ages 18–29 own a tablet (American Press Institute, 2014).

Since 2005, social media use has increased rapidly and offers users new and different ways to form social ties and new means to consume and produce information and knowledge. By 2015, 90% of young adults ages 18–29 were users of Facebook, Twitter, or LinkedIn (Perrin, 2015). Though Facebook is still the most popular of social networking sites for this age group (88%), the popularity of Instagram (59%) and Twitter (36%) is also evident in this age demographic (Greenwood, Perrin, & Duggan, 2016). Twitter and Instagram appear to hold particular pull among Latino/a and African American users. The Pew Research Internet Project noted that African Americans and Latino/as use Twitter and Instagram roughly twice as much as Whites. Thirty-four percent of African Americans and 23% of Latino/as use Twitter's and Instagram's quick and short text, photo, or video information bursts. Also popular among college students is the chat app Snapchat. A self-destructing photo and video-sharing app, Snapchat appears to appeal to users under 25 years of age. On campus, 77% of college student users report using Snapchat every day. Women college students report sharing selfies (77%) more than college men (50%) (Duggan et al., 2015).

Like race, ethnicity, age, and gender, a user's income level also impacts use in particular ways. Interestingly, lower-income earners are more likely to use Facebook than higher-income earners. Greenwood, Perrin, and Duggan's (2016) data on technology use and income level show that 84% of adults earning below $30,000 use Facebook, while only 77% of adults earning $75,000 and above do so. Adult Facebook users earning $30,000 or below are more likely to be African American or Hispanic. Duggan and Brenner (2013) reported that cellphone use across income brackets is fairly consistent, however. Eighty-five percent to 97% of all adults ranging in income brackets from under $30,000 to over $75,000 are cellphone users. Tablet use, however, does seem to be a phenomenon of higher-income and not lower-income earners. Only 26% of adults who earn less than $30,000 own a tablet, while half of all adults earning over $75,000 own a tablet device (Duggan & Brenner, 2013).

Trends in social media, social networking, and mobile technology use among college students frequently reflect changes that occur in these technologies while students are in high school and how their use changes while in college. These changes in use often reveal developmental and situational needs. Online social networking was central in the lives of today's college students as high school teens. In 2007, Pew reported that teens ages 12–17 (covering many of today's college students) were increasingly using Web 2.0 technologies to post images and videos that often received feedback from friends. Texting, whether on a cellphone or on social network sites, was increasingly the means for contact with friends rather than email and landline phones. Messaging on social media and texting on cellphones enabled college students as teens to make more frequent and easier contact with friends. By 2013, 82% of teens ages 14–17 were on either Twitter or Facebook, but using these sites for decidedly different purposes (Duggan & Smith, 2013). These older precollege-aged teens began to differentiate their social media site use, using Twitter, Instagram, and Snapchat as sites with lower social and relational stakes. As quick self-expressions, Twitter, Instagram, and Snapchat do not require investing too much in impression management and relationship building (Martínez Alemán & Wartman, 2009). Because they do not require premeditation and planning, these sites require less deliberation. Additionally, especially for precollege-aged teens, these sites can be hidden from parents and other adults with untraceable, creative aliases. In contrast to Facebook, older teens perceive these sites as less likely to be relationally risky. Instead, they use these sites for self-expression and sharing, and to follow news, celebrity gossip, and pop culture. It seems that older teens (many of whom are likely now college students) intentionally use different social media sites for specific purposes.

A Trio of Social Media Sites at Millbrook

INTERVIEWER: Is social media big at Millbrook?

CAROLINA: Social media is huge at Millbrook . . . Facebook, Twitter, Instagram . . .

In almost all cases of students who used social media in this study, FGCS regularly employed more than one online platform in their daily usage. These platforms varied from longstanding media-sharing applications such as YouTube to popularizing messaging apps such as Snapchat, but the majority of students, like Carolina above, cited a trio of social media sites that were most dominant at Millbrook: Facebook, Instagram, and Twitter.

FACEBOOK

Facebook was unquestionably the dominant social media on campus. As previous reflections and findings have specified, students described Facebook as the centralized online forum that encompassed student life on campus. As Nico further affirmed, "Everyone on campus does a lot of Facebook stuff. People will even change their profile picture to event flyers. Then you'll always get event invites on Facebook. Facebook is big."

Facebook was often referred to as the unofficial forum and source for FGCS to acquire information regarding the intricacies of student life on campus. Mason stated, "I've been more dependent on Facebook because everything is posted on Facebook. Like club events and announcements. I have to use Facebook a lot more often because of it." In an example that describes Facebook's multifaceted capabilities as an information-dispensing and -acquisition center, Meena described how "there is a special group on Facebook now called the 'Millbrook Free and Sale Group,' so if you're looking for something or wanting to sell something, you'll typically post on that page." As Meena described, Facebook facilitates multidirectional streams of information distribution. Students agreed that Facebook provided opportunities to simultaneously learn about campus life and contribute to the social environment of Millbrook.

The accessibility of Facebook as a compatible mobile application also proved to be advantageous for students' involvement online, as FGCS were consistently able to access this social media at any point and time throughout their time on or off campus. Additionally, Facebook events possessed the ability to automatically sync to cell phone and tablet calendars, ensuring students received notification of upcoming program invitations and campus commitments. This mobile accessibility permitted FGCS to access the virtual hub of student life at almost any point and time at Millbrook. When asked on average how often he checks his social media on his phone or iPad, Gabriel stated, "All the time. Anytime I'm not doing anything. Whenever I'm bored or waiting for someone, I just go on Facebook." Such behaviors indicate a dependency on technological devices, and further validate the characterization of today's students as being citizens within both the physical and the virtual campus communities.

INSTAGRAM

> Instagram, so much Instagram, so bad. I'm just like okay, okay, like every five minutes. Like why? I don't know.
>
> —Yvette

Yvette's quote reflects the sentiments of the majority of her FGCS peers that the photo-sharing app Instagram was the second most popular social media at Millbrook and was lauded by FGCS for its ability to capture and convey specific aspects of campus life that reflected an individual's creative perspective. The ability to document campus experiences, programs, and sites through personal photography on Instagram allowed students to visually share and document their experiences on campus with their peers. Meena explained the appeal of this social media app: "I use Instagram a lot because I love to take pictures. That's just another way to connect with friends, but I also use Instagram as my own personal photo album to just show off things."

Andrew's observations of social media use at Millbrook also support the fact that Instagram is largely embraced by the student culture. He described how students regularly take photos of picturesque buildings on campus and often tag them online. "People post pictures of Founders Hall on Instagram. That's a thing now. They take photos in different filters, tag it with #FoundersGram. I don't do it, but I see it on my Instagram all the time." Nico confirmed the student culture of capturing campus buildings on Instagram: "I feel like people use Instagram a lot at Millbrook. They take pictures of themselves all the time, and of Founders Hall, and the quad. They take pictures of everything, I think."

Although not as pervasive as Facebook, Instagram was still widely used among the student body and FGCS at Millbrook, and we saw an increase in use over the years of our study. According to Ruby, an EOP peer mentor who assessed the state of social media use among the students in this study during the EOP, "Instagram was all over. I saw Instagrams about the talent show, the competitions, and the trips. I think Instagram is a big connector between all of the EOP students." When asked about the appeal of Instagram as a social media site, students often replied that the app was "fun" or a "creative way to express yourself." Andrew specifically characterized Instagram as an outlet to "show your artsy pictures and show people special moments. It's a just a neat way to post something cool." On the other hand, Mason viewed Instagram as an expressive outlet and as an effective way to see into the world of his friends. He explained: "I use Instagram out of like my own uniqueness and out of like wanting to see what other people are doing. I pretty much use it to keep up-to-date with other friends and see what they're up to. A lot of my friends like to take pictures and stuff like that, so it's kind of cool to see where people are venturing off to places."

Students were notably more active on Instagram than on Facebook in constructing posts that reflected their own personal perspectives of student life. Whereas Facebook predominantly served a utilitarian role of a campus directory for peers and activities, students took increased ownership on Instagram to personalize their photos, which often reflected memorable life events or were meant to inspire others. As Gabriel described: "Instagram is good to see the experience that others are having on campus. I guess it can even lift you up. If you see someone doing something positive or smiling or doing something funny on Instagram, it's going to make your day better." The personal nature of Instagram, coupled with the possibilities to be creative on the app while staying connected with friends, were common features cited by students as reasons the social media site was beneficial to use on a regular basis.

Students stated they spent a considerable amount of time on Instagram. When asked to elaborate how often she was on the site, Carolina was unable to specify an exact amount of time but stated, "A lot. Every time I'm on my phone, I'm checking Instagram." The regular use of and connection to Instagram was identified by the app's ability to be conveniently accessed on mobile devices and its ability to synchronize activity with Facebook accounts. As Ava described, "People link Instagram to their Facebook, so you can share events across them both. . . . And everyone wants to take photos these days. Everyone has an iPhone now." The constant and seamless connections to this app further embody how social media fully immerse both student culture and social life at Millbrook.

TWITTER

Twitter was cited as the third most common social media platform at Millbrook. Students in this study described Twitter as a popular social media app because of its characterization as a fast and efficient information acquisition site. Krista explained: "Twitter is where I can get on and read something really quick. It's not something I actually have to take time out for." While the social media site maintains mobile and accessible functionality for messaging, tweeting, and photo sharing, students rarely engaged in these interactions on the site. Rather, students were passive observers who preferred to acquire information on this social media and characterized Twitter as the forum that was most pragmatic for reading breaking news, browsing developments in politics, and catching up on world events. Camila elaborated on her use of Twitter: "I use Twitter a lot because I found out, it's like the news. You find out so much stuff is going on, like what's trending and politics, anything on Twitter."

In other instances, Twitter was cited as the social media that was most suitable as a virtual proxy for connection with celebrities and a clearinghouse for developments in pop culture. Andrew noted why he uses Twitter: "I follow so many hockey analyst guys on ESPN, Canadian guys, TSN, ESPN, CBC, NBC, all of them. Or, if I hear there's a news story somewhere, I just go straight to Twitter. Like when Stephen Colbert was named to take over for David Letterman. I just went straight to Twitter to see what was there."

Overall, students maintained that Twitter primarily served to externally connect them with news and pop culture, but rarely to their strong and weak ties at home or at Millbrook. Twitter was seldom used to connect with clubs or campus resources, and students typically did not cite this social media site as an effective forum to advance and strengthen relationships with peers on campus. These descriptions of Twitter's use on campus reflect the primarily impersonal and information acquisition-oriented nature of this social media for FGCS at Millbrook.

The previously documented testimonials and examples of the influence Facebook, Instagram, and Twitter had on students in this study further reinforce the interconnected nature of student life with social media at Millbrook for FGCS. Facebook continued to be the most centralized forum of campus life that promoted involvement in the campus environment. Campus clubs consistently and unrelentingly leverage the multiple marketing capabilities of Facebook to draw publicity for individual meetings, or create hype for signature events. FGCS regularly and naturally turned to Facebook groups and their peers on social media over administrative offices to seek insights that best guide students through the social setting of campus life. Students also creatively captured their campus memories through photography on Instagram, acquired news while connecting to pop culture celebrities through Twitter, and seamlessly engaged on all of these platforms. The totality of these connections on social media undoubtedly captures how student life is shaped, is captured, and now co-exists as a vibrant virtual community alongside the traditional campus setting.

Students' Readiness for Technologies on Campus

A more focused view of college students' technology use reveals trends in academic applications. The EDUCAUSE Center for Applied Research's Study of Undergraduate Students and Information Technology's longitudinal data offer insight into the state of students' technology use in a given year as well as exposing tendencies and trends. The 2013 EDUCAUSE report presents data that support the notion that social media use in higher education has changed with a great deal of speed over the past decade (Dahlstrom, Walker, & Dziuban, 2013). College students are no longer hesitant to engage with blended-learning instructional methods, and those courses that include work both in and outside the classroom, delivered online and offline, are in fact part of the expectation among students. Concurrently, students are increasingly equipped with mobile, online-ready devices, which has allowed for a transition into the blended-learning curriculum. Although some students still claim a preference for computers over mobile devices due in part to larger screen size, easy-to-use keyboards, and simplicity of use, mobile devices are now seen as increasingly acceptable among undergraduates (Dahlstrom, 2012; Dahlstrom, Walker, & Dziuban, 2013).

The most recent data on undergraduates' academic use of digital technology indicate a willingness among students to embrace mobile and online media as academic tools but also a desire to have training when it comes to these new tools (Dahlstrom, 2012; Pearson Foundation, 2012). Overall, students today express a sentiment that it is clear that technology helps them to achieve their academic goals, but when it comes to the application of technology tools within coursework, students want their instructors to personally clarify expectations and benefits of use (Dahlstrom, Walker, & Dziuban, 2013). Early studies within education indicate that mobile technology in particular can have a positive impact on student achievement, learning, and engagement, primarily due in part to key subject-focused applications (Heinrich, 2012; Manuguerra & Petocz, 2011; Murphy, 2011).

The parallel rises of university access to and implementation of technologies and the growing comfort of students with technology-based academic tools present a host of opportunities for higher education. Students report overall positive feelings toward faculty use of mixed media within coursework, and the technology to offer new forms of instruction is becoming more accessible to institutions as prices fall. Further, the rise of social media has allowed for technology to aid the building of bridging and bonding ties between student populations, as well as the development of ties between students and campus resources and staff (Dahlstrom, 2012; Martínez Alemán & Wartman, 2009). Student use of social media is also positively correlated with various traditional undergraduate engagement measures (Chen, Lambert, & Guidry, 2010; Heiberger & Harper, 2008; Junco, 2011).

But because college students use technologies selectively and for particular purposes, when and how college students will use Web 2.0 technologies is a point of concern for higher education. Context is important for today's undergraduate student user, as the online spaces in which interactions occur help dictate the people with whom the students are willing to communicate and how that communication is enacted. The majority of undergraduate students are opposed to blending their social and academic lives online. Although college students feel a great deal of ease in connecting with peers online, they expressed discomfort when presented with the idea of using social media to communicate with faculty members (Dahlstrom, 2012). Consequently, colleges and universities must consider the suitability of technologies for desired learning and developmental outcomes. Separation of social and academic lives remains a priority for today's students, and this sentiment creates a limitation for Web 2.0 technology use by colleges and universities.

Despite students' regulation of their technology use, colleges and universities continue to explore and examine the utility of social media and mobile technologies for student developmental and academic outcomes. Increasingly, the role of social media within the higher education environment is thought to be essential to impacting student outcomes. In particular, social media applications are now seen as tools that can aid undergraduate engagement (Heiberger & Harper, 2008; Junco & Cole-Avent, 2008; Violino, 2009). As seen above, these applications are such a part of the student experience and so seamlessly integrated into the daily life of a college student that higher education researchers and practitioners argue that these media must be used to improve college student engagement (Junco & Cole-Avent, 2008). In response to these calls, administrators have added full-time staff members to work exclusively with social media (Violino, 2009). Efforts made by higher education leaders to develop social media strategies have yielded positive social capital benefits for college students, who are now able to easily access campus resources online (Benson, Filippaios, & Morgan, 2010; Heiberger & Harper, 2008; Junco & Cole-Avent, 2008; Violino, 2009).

Scholarly research in the field of higher education has explored the use of social media on campus. For example, current research has investigated the role of social media in building relationships that are crucial to social engagements and community building (Morris et al., 2009; Yazedijian et al., 2008) and in creating contexts that encourage contribution and collaboration among students (Muňoz & Strotmeyer, 2010). According to Muňoz and Strotmeyer (2010), social media are an ideal medium to facilitate engagement by integrating relevant campus information with social networking sites used by students for personal reasons. Junco (2011) examined the relationship between Facebook use and traditional benchmarks of student engagement, concluding that Facebook activities were strongly predictive of engagement. Chen, Lambert, and Guidry (2010) also used the National Study of Student Engagement (NSSE) items and measures to assess the impact of online technology on engagement, finding a positive relationship between use and engagement. The same can be said of Heiberger and Harper's (2008) study of college student engagement and Facebook use, and the Higher Education Research Institute's study of first-year students and online technology use by Saenz et al. (2007). In each of these studies, researchers inferred a positive correlation between online social networking and traditional college student engagement. Data from the Pew Internet and American Life Project (Hampton et al., 2012) suggest that social ties made through Facebook are associated with higher levels of social support, implying that online social networking can provide the relational space in which forms of social capital can be transmitted.

When integrating social media strategies into their work, administrators must take into account the different usage patterns based on student subpopulations at the institution. Today, African American online users are on Twitter in significantly higher numbers than White online users (26% versus 14%, respectively; Duggan & Brenner, 2013). Latino/as, like African Americans, find Instagram more appealing than Whites (Duggan et al., 2015). Consequently, because students use social media and apps for particular purposes and social identities and position like race and ethnicity can affect use, college and universities face a unique challenge if they are to leverage the power of these Web 2.0 technologies to improve student outcomes. It seems that the purposeful leveraging of social media by colleges and universities requires identifying the connections between the social context of users and the particular media's utility for students. In particular, the needs of racial and ethnic and other underrepresented students on campus could be addressed with technology if these considerations are taken into account. Although social media trends are just beginning to be explored within academia, it is vital that researchers and practitioners not view users as uniform. As social media develop and user behaviors adapt, individual and niche trends must be observed to fully serve all college communities.

GOOD IN THEORY—BUT NOT THERE YET IN PRACTICE

While students are more open to embracing social media use in some ways for engagement, and the research outlined above shows a benefit for the use of social media by administrators, there can be an administrative aversion to social media, as was seen at Millbrook. Given their steady connection to Facebook, FGCS recommended campus administrators who were interested in further engaging students in campus life to more intentionally utilize social media as a means of outreach for publicizing events and for engaging with students. Yvette recommended administrators use Facebook event invitations to remind students of an office's events: "If we're going to have an EOP reunion, it would be helpful if there was a Facebook event for that instead of an email, because you see, Facebook goes on my phone calendar." Another student specified that the newsfeed on Facebook typically is more effective in capturing students' attention to upcoming events than departmental invitations sent by email. Yvette conveyed her suggestion to administrators: "I guess it would be more helpful to see events [that administrators want to see us attend] on my newsfeed. Like a list of the events that are coming up on my feed would be helpful so we don't have to go through our emails."

Unfortunately, those who work on campus typically rejected the suggestions that social media should be used by campus administrators as frequently as students would recommend. Mark, the administrator who oversaw the EOP, reflected on his role in using social media to engage students in this program: "I myself don't have Facebook and so I'm not involved with any kind of Facebooking. I'm probably the worst person to ask. I don't know anything about Facebook. I likely wouldn't get an account either because I'm not putting my personal info for students to see online. I definitely see that being my big concern about using social media."

Ironically, while this administrator stated his discomfort using social media, the EOP office maintained a specific Facebook group that it hoped would facilitate connections and engagement between FGCS in this study. Unsurprisingly, students largely ignored this page's presence due to the office's inability to understand how to manage this social media and how to intentionally engage users online. The EOP essentially assumed social media did not require specific purposing and attention for student use, which was a detriment to the success of the Facebook group.

This lack of social media literacy and absence of intentional management of a Facebook group by campus administrators further threatens the promotion of advancing campus capital online.

The intersection of social media with campus life has the potential to provide robust dissemination of campus capital through strategic engagement of FGCS, but this possibility is threatened when campus administrators fail to maintain competencies that can positively advance these outcomes. While peers are prominent connections to facilitate campus capital, administrators can play an equal, if not more crucial role in advancing knowledge, skills, and resources to FGCS that ensure successful transition and incorporation into the social life of campus. Students continually engage with social media as a lynchpin of campus life, and continued administrative disregard for proficient engagement on sites such as Facebook further perpetuates gaps in campus capital that hinder FGCS in their transitions into campus life.

MOBILE TECHNOLOGIES AND PEERS

Research on student use of technology indicates an opportunity for institutions to utilize undergraduate peers as key connectors between universities and both current and incoming students (Benson, Filippaios, & Morgan, 2010; Martínez Alemán & Wartman, 2009). Many students use social media as their primary source of information regarding a college campus upon admission. For these enrollees, for whom socialization to the university community is a process that occurs largely online and before arrival, peer student leaders can play an essential role in the development process (Martínez Alemán & Wartman, 2009). On social networking sites like Facebook, student leaders can act as "conduits" of social capital on campus and can serve as "cultural translators" of campus social codes and norms (p. 94). As Martínez Alemán and Wartman (2009) demonstrated, college students use social media to communicate identity and decipher campus culture from their specific developmental, gendered, classed, and raced subject positions.

In alignment with literature cited above, from the EOP peer mentors to classmates and campus organization colleagues, our FGCS rely on their peer networks to acquire information related to academic processes and services. Communication about everything from which professor to take for a particular class to coordination of group projects occurs both in person and through social media. Although students sometimes use email to contact classmates they do not know or for certain aspects of student organization business, more informal modes are generally preferred for peer-to-peer communication outside a face-to-face (F2F) context.

For learning or asking advice about important academic-related information, such as when to register for classes or how to go about studying abroad, students indicated that they rely on campus authorities or trusted individuals within their already-established social groups, especially the upperclassmen they know. With the exception of official emails from administrative offices or advisors or information posted on office websites, most of these conversations occur in person rather than online. Nico mentioned that he had learned about internships and other things to expect in college from the seniors in the men's discussion group of which he was a part. Similarly, Andrew discussed what courses to pick for his class schedule with peers in his dance group "because there are four freshmen and three sophomores and one senior, so these guys definitely know how this thing works." Yvette spoke about how the juniors and seniors on her Latin dance team helped her with the classes they had already taken.

Although students seem to learn about academic programs or resources mostly through flyers, official email communication, or word of mouth, they occasionally obtain useful information through social media, though not with the same frequency as information about social events. One student referred to a Facebook page that served as an online classifieds site: "I know there's like a selling sort of page that you can sell your textbooks, furniture, anything. That was really good first semester when I was trying to get books because they're really expensive at the bookstore." For keeping track of Career Center events, such as resume and cover letter review sessions or job fairs, Meena recommended using the app version of CareerCentral, the university's career portal. These instances, however, seemed to be the exception to the rule with regard to students' methods for obtaining information relevant to the academic aspects of their college experience.

Web 2.0 Technologies in the College Classroom

> I think the iPad training definitely helped. It's funny, because the example I always think of is last year when Ana was observing my class and [my iPad] kept going to the screen saver, I didn't even know there was a way to turn that off. I thought that's what iPads must do, and so this year having somebody show me how to do that, that just made it seem like okay [I can do this].
>
> —Martha

> My [one] teacher, she doesn't really like the technology. She doesn't like it because everyone's always coming in the class on their phone, and they don't talk to each other, they're on their iPhone, or iPad, or laptop. So she's completely just like no technology, talk to each other, these are your classmates. So it had nothing to do with classes and the notes, and what's going on, but she just wanted us to socialize and talk to each other. So she banned all technology.
>
> —Ava

The theoretical and empirical scholarship on Web 2.0 technologies is quickly evolving. In particular, the collaborative yet autonomous nature of learning that these media and mobile platforms encourage makes them uniquely suited as a means for college and university instruction and student learning. Raine and Wellman (2012) argued that online social media are the "new networked operating system" (p. 9) that can increase opportunities for learning, problem solving, decision-making, and personal interaction. Though they acknowledge that a user's preexisting social capital (e.g., what they know, how they know it) impacts the new capital that can be acquired, it is still the case that Web 2.0 technologies increase a user's learning prospects. Social media, then, can present users with more occasions to direct or produce their own learning, as well as consume knowledge (capital) that is communicated by other users.

As environments that are both personalized and arguably universal, social media have proven effective frameworks for supporting informal and formal learning in college (Dabbagh & Kitsantas,

2011). Social media enable students to manage and maintain spaces online that connect them to peer and other learning networks on and off campus and that complement and enhance traditional learning approaches (Alexander, 2006; Dabbagh & Reo, 2011; Horgen & Olsen, 2014; McLoughlin & Lee, 2010). Analyses of classrooms employing social media have demonstrated that students take on more responsibility for their learning while faculty facilitate learning (Svendsen & Mondahl, 2013). Despite concerns and shortcomings, combining social media with other forms of instruction has shown positive effects in the college classroom (Acar, 2013; Muñoz & Towner, 2010; Towner & Muñoz, 2011). Given that social media now shape and regulate campus culture (Martínez Alemán & Wartman, 2009), it is not surprising that social media are quickly becoming part of instruction on campus.

As instructional technology, social media and other Web 2.0 technologies can prove useful for traditional brick-and-mortar higher education. First and foremost, the nature of social media technologies affords the user a great deal of autonomy. User personalization characterizes these sites and when placed in a learning context, students can engage in self-directed learning that is customized to their particular needs. The guiding principle and spirit of social media is their capacity to provide individual users with opportunities for self-direction and independent learning, even given their collaborative character.

Faculty who incorporate Twitter as an instructional technology report greater and better student participation in these classrooms, suggesting that as a platform it motivates engagement in learning. We know that the relationship between students' motivation and learning is salient, and that performance differs between students with different motivational conditions (Liu, Bridgeman, & Alder, 2012). Consequently, instructional technologies that either increase student motivation or help faculty gauge student motivation can be valuable pedagogically. Early research on the relationship between student motivation and instructional social media suggests that these technologies can positively impact student engagement (a proxy for motivation) and grades (Junco, Heiberger, & Loken, 2011).

For faculty, social media and other Web 2.0 technologies provide a means to additional layers of assessment. Through Twitter, for example, faculty can create backchannels of student communication that can give instructors a better sense of student comprehension and class climate (Greenhow & Gleason, 2012). Twitter feeds can provide faculty more avenues for more frequent evaluation of student understanding, as well as additional ways of measuring student input. For example, if tweeting on particular points of the lecture is required, faculty can weigh the strength or substance of the tweet and grade accordingly. This is especially beneficial for students who are reticent participants in class (Martínez Alemán, 2014). Through the implementation of Twitter, faculty can also give students quicker and more frequent feedback (Taylor, 2010). But perhaps more importantly, by using social media and other Web 2.0 technologies faculty themselves become more active participants in student learning. As researchers have observed, faculty who use Web 2.0 technologies increase their own participation and become much more active partners in the teaching-learning relationship.

KEY CLASSMATE CONNECTIONS

The social character of Web 2.0 media brings to college classroom instruction many opportunities for collaborative learning. Not only do these media extend teaching and learning beyond the classroom and its restricted meeting time, but also they create the conditions for expanded peer learning and informal learning. Social media are especially convenient and advantageous for facilitating cooperative learning among students. Undergraduate students can utilize the varied functions of these media to deepen and broaden meaning and to create new funds of knowledge. Twitter-friendly classrooms enable students to hear different points of view and to learn more about their classmates. Through their academic engagement with peers on these technologies, students are also circulating course knowledge and other academic capital (Valenzuela, Park, & Kee, 2009). When faculty utilize social media as a part of their instructional strategy, they find that both formal and informal learning is positively affected. Social media appear to facilitate informal learning and to enrich class discussion and broaden the connections between concepts (Junco, Heiberger, & Loken, 2011). In institutions that serve primarily low-income FGCS who are often English-language learners, the integration of various technologies in coursework has generally had positive effects on students' classroom experiences. Low-income, language-minority FGCS have found assignments more interesting with technology, and have experienced more discussion and interaction with classmates when technology is integrated in coursework (Razfar, 2008).

In addition to acquiring academic information from their peers, our FGCS also learned new skills or technologies from each other. Justifying his recommendation that the university offer more formal tutorials for students to learn how to use technology more effectively, Gabriel imparted that he had "learned a lot about, I guess, the little tricks to use the iPad from people around me, whereas if I just had learned that from the beginning I probably would have used it more quickly." Nonetheless, students did indicate that transfer of knowledge about technology occurs, and this seemed especially true during the EOP. For example, a presentation assignment from Martha, one of the English instructors, resulted in students learning how to use Prezi, a dynamic presentation software program. She reflected on how this knowledge benefited both her students' quality of work and their self-confidence:

> I had them do these presentations, and the first student or two used Prezi, and then some of the other students really liked it, so they wound up sort of teaching each other how to use it. And they did these beautiful, totally impressive presentations, and I wouldn't have been able to show them how to do that. I don't know how to use Prezi, but I think it gave them a real sense of mastery of this technology that they could put together something that looked really good.

Another useful tool that some students learned about from their peers during the EOP was GroupMe, a versatile group text messaging app. Ava indicated that until one of the peer mentors added their entire group to a GroupMe conversation to coordinate meeting times and send important reminders, she had not known what GroupMe was. Considering the frequency with which students mentioned using this app to communicate with their college friends and acquaintances, familiarity with it seemed to constitute an important piece of campus capital.

Although references to GroupMe usually occurred in the context of conversations about social interactions, students also mentioned the app as helpful in communicating with classmates about group projects. Carolina explained: "Basically just somebody will add everybody's names on one group, and then we'll just talk about it, like where we're going to meet, what are we going to, if we like something, and just, it's very helpful." Along with GroupMe, Google applications and Facebook were frequently mentioned tools for connecting with classmates.

Students most commonly utilized Google applications (e.g., Docs, Slides) for the sharing functionality to easily collaborate on group assignments and presentations. Some, such as Krista, also found other creative applications for Google Docs: "There's like eight of us, because the first time we met, it was just like individual notes . . . and so after the first exam, we just decided to just make a Google Doc in the beginning and then as we read and as we take notes in class, everyone can just go on, or like one or two people will go on and do, fill it out and stuff like that." With Facebook, students frequently connected to request information, such as class notes or clarification about an assignment, from individual classmates, and the Messenger feature was also sometimes used for group chats to coordinate shared projects or study groups.

In rare cases, instructors created a Facebook group or page in an attempt to intentionally foster such interactions. Martha's class page provided a space for students not only to ask questions but also to respond to each other's questions. By structuring the page as a communications tool as well as a platform for supplementing the course material, the professor effectively accessed and facilitated her students' natural modalities for participating in this out-of-class interface. However, in contrast to student-created

Facebook pages or groups, such faculty-created pages typically seem to possess a limited lifespan. In attempting to explain why the page her professor created saw no traffic after the course ended, Carolina hypothesized: "I think pages really work for homework, or like groups, or projects, stuff like that, more academic based than socially. Socially you just want to stay with your friends, not with like, unless you make friends in the classes. I think it's mainly used for academics. So like questions or homework assignments." Other students also characterized interactions with classmates as generally more mission-oriented and related to coursework. If a prior or external association did not exist, the relationship could, however, progress to friendship if the students shared a great conversation in class or worked closely together on a group project, for example. The relationship might also evolve if, after establishing a social media connection, they started to "follow" each other, thus remaining linked outside and after the class. Thus, while most classmates remained weak ties, sources of course-related information and assistance, social media could serve to strengthen these ties once an initial connection had been established.

Finally, the technologies the students used were in and of themselves a source of campus capital, especially within their peer networks. Particularly for the students who could not financially justify the expense of a laptop, the iPads that were provided to the EOP participants significantly leveled the academic playing field. Rather than having to rely on communal desktop computers or temporarily checking out a laptop from the library, students could use their iPads to access online textbooks and PDF readings, view PowerPoint slides, check the course management system for class information, and complete assignments, among other academic tasks. One of our key findings was that students needed to be able to keep the iPads for more than one year. In the first year of the project students were told that they needed to return the iPads at the end of the year—and some students reported that they did not want to invest in learning this technology that was only temporary. We saw a big difference with our year 2 cohort, who knew from the beginning that they would be able to keep their iPad for all four years if they remained participants in our study. In addition to offering academic benefits, possession of the iPads engendered a greater sense of belonging at an institution with a campus culture of privilege.

CHALLENGES TO INTEGRATION IN THE CLASSROOM

Faculty have different attitudes about technology use in the classroom. As is the case in the general population, more than 80% of faculty are social media users, and at least 45% use these sites for professional purposes on a monthly basis. Recognizing that the interactive nature of social media and mobile technologies can produce the conditions for better learning, a growing percentage (34%) of faculty incorporate these media into their teaching. Faculty are also cognizant of the fact that digital communication has increased their contact with students. Despite faculty's increased use of social media and other Web 2.0 technologies, they still tend to rely on consumptive media like videos (YouTube) and podcasts and have reservations about investing their time to learn these new forms of instructional technologies (Moran, Seaman, & Tinti-Kane, 2012). Many faculty are unconvinced that there is enough evidence that social media use is positively associated with learning outcomes (Moran, Seaman, & Tinti-Kane, 2012). In contrast, their students are more open to using social media to support coursework (Roblyer et al., 2010), though they, too, must perceive direct relevance between the media and course objectives before they adopt their use (Acar, 2013).

Further restricting the use of social media as an instructional tool are students' and faculty's concerns for privacy, faculty's resistance to dedicate time to learning these instructional technologies, and the identification of this space as strictly social by students (Cao, Ajjan, & Hong, 2013; Martínez Alemán, 2014; Martínez Alemán & Wartman, 2009). Students and faculty still fear the loss of privacy on social media despite increased knowledge of privacy settings and restrictions. Students and faculty alike worry that private matters posted on social media could become part of the academic experience, breaching the unspoken divide between their personal and academic lives. Data indicate that faculty continue to use sites like Facebook mainly for personal postings and communication, consequently framing that space as private (Moran, Seaman, & Tinti-Kane, 2012). Like many older adult users, faculty understand social media sites like Facebook as spaces for communication with friends and family. Though more and more faculty report using Facebook and other social media for professional purposes, these are communications with peers and not beyond the scope of their private networks. Incorporating students in these spaces disturbs faculty's construction of privacy and interrupts their online social niche.

Faculty's privacy concerns regarding the use of social media as instructional technology are mirrored in students' perceptions of these spaces. Because students believe social media to constitute spaces free of adult supervision and authority, positioning social media as a space in which faculty can intrude and impose judgment is undesirable. Students still perceive social media as social space in which there should be limited intrusion of authorities. Recent data on teen use, for example, reflect this concern. Teens often have two accounts on such sites as Facebook to steer adults and adult authorities to the profile swept clean of any inappropriate material and friends to the profile that is more in line with their social norms (Pew Internet and American Life Project, 2013). When students participate in course-related communication with faculty on social media it comes as no surprise then that they do so reluctantly (Moran, Seaman, & Tinti-Kane, 2012).

The micro-blogging site Twitter seems less problematic than Facebook for faculty. Not surprisingly, Twitter has proven to have positive effects on peer learning and students' language skills. Students find that reading each other's micro-blogs is helpful, that peer feedback improves their peer communication, and that this communication stimulates reflection (Ellison & Wu, 2008). Twitter as instructional technology appears to improve students' comfort with classmates and seems to engage students in classwork (Junco, Heiberger, & Loken, 2011; Smith & Tirumala, 2012). But as instructional social media, Twitter use has also been shown to increase students' communication and consequently their interactions with faculty (Greenhow & Gleason, 2012). The nature of these interactions is such that they are less likely to intrude on faculty's privacy. Unlike communications on Facebook, Twitter communication between faculty and students appears limited to course content and academic material. As instructional media, Twitter is not perceived by faculty as personally intrusive perhaps because as a micro-blog, Twitter lends itself to intellectual content in a way that Facebook does not. Tweeting appears more an act of brainstorming, as a way to extend class discussion in and outside the classroom, and as a method of scholarly collaboration between peers as well as between faculty and students.

The ease of use and perceived relevance of social media and Web 2.0 technologies to course objectives appear to be the lynchpins in faculty and students' adoption of social media for academic purposes. As Kaufer et al. (2011) observed, when the nature of social media corresponds to course objectives, student learning becomes multimodal and productive. Kaufer and colleagues' examination of social media in a writing classroom illustrates how peer social networks can develop as a means to deepen students' understanding of texts and themselves as readers, for example. Texts can be probed by annotating with peers on social media, also helping students to know each other as readers and authors, and to create communities of learning. The classroom writing salon is possible because faculty understand the pedagogical value of collaboration and social media and students see the relevance of the technology for course objectives (Kaufer et al., 2011; Moran, Seaman, & Tinti-Kane, 2012).

The research and scholarly literature on the value of tablet technology and social media for the acquisition of social capital was presented as a necessary part of the foundation for the intervention presented in this book. The impact of tablet technology (in particular, iPad technology) on college student development, college student learning, and college student engagement has yet to be fully examined by researchers, but early work points to the value of iPads as complementing and accompanying academic work and as a means for students to extend their informational reach quickly and easily. As a mobile technology, iPads and smartphones appear to provide FGCS with flexibility and improved access to more sources of campus capital.

Social media are predicated on the actuality that our F2F social networks can be leveraged to advance our social capital and that this form (and new forms) of social capital acquisition is possible through online relational ties. Scholars and researchers of online social networking point to the many different ways in which social networks on social media impact the acquisition of varied forms of social capital generally, and campus capital in particular. Research on these media reveals the complexity, nuances, and variability in campus capital acquisition by students across embedded identities such as gender, race/ethnicity, and first-generation college-going status. The distinctions between insider and outsider campus capital suggest that the relationships between social media users and the bonds that connect them determine the type of capital circulated and its saliency. Students' network of friends on Facebook, for example, is a web of relationships in which bonds have varying degrees of trust and connectivity and instrumental value. In all, however, the research and higher educational communities now acknowledge social media as a conduit for campus capital.

4
Transition and Campus Engagement

WITH CONTRIBUTIONS BY KEVIN GIN & SCOTT RADIMER

Camila walks through the clubs' fair and first-year BBQ during Millbrook University's fall semester's welcome week activities. The fair is busy with hundreds of students gathered on the campus quad for the afternoon festivities. Student leaders from each of the organizations stand in front of hand-drawn signs advertising the date and time of the first club meetings of the semester, while other club members hand out candy, encouraging new students to sign up on mailing sheets. As Camila walks through the fair, she feels more and more anxious. She cannot find the one club she most wants to get involved with this year, the Millbrook Dance Ensemble. As she reaches the final table at the fair, she realizes the Dance Ensemble did not show up today. She thinks to herself, "How will I make friends this semester if I don't find someone else who shares my interests? Do I really have to join these other clubs if I can't find the club I want to get involved with? And will getting involved on campus take me away from the time that I could be spending each night connecting with my family back home?" Feeling the looming fate that she will not leave today with information about how to join the one club she wants to be involved with the most, Camila leaves the fair and sits down by herself at a picnic table at the campus BBQ.

Feeling disappointed she could not find the Dance Ensemble, and feeling lonely because she has no one on campus to commiserate with in this moment, Camila decides to pass the lunch hour by taking out her iPad and scrolling through her Facebook. She is greeted with three new notifications on her profile page. The first two notifications tell her two of her peer mentors from the Educational Opportunity Program (EOP) have changed their profile pictures to a club flyer advertising the Dance Ensemble's first informational session of the year, which will take place this evening at 7:00. Camila is at once relieved and excited. She quickly notes the time and classroom location of the meeting in her iPad calendar, feeling hopeful about the prospect of meeting new friends on campus.

Camila then turns back to the other notification on her Facebook showing that she has been tagged in a picture that was posted online by her younger teenage cousin Sydney. The photo is of Camila and Sydney the week before Camila moved to Millbrook and the caption reads "Miss and love you, Cuz!" Camila instantly feels homesick, but sees Sydney is online and sends a greeting through Facebook Messenger. The two chat for a few minutes, and Camila tells Sydney about the Dance Ensemble's meeting and her growing excitement. The two cousins virtually chat for a few minutes about each other's weekends and how life is back at home. Camila eventually signs off, telling Sydney how much she misses and loves everyone back home. But before Camila walks back to her room, she realizes she wants to do one more thing. Camila goes to her account on Facebook and changes her profile picture to the Dance Ensemble's flyer that she saw on her peer mentors' pages. "Maybe someone else I know was also looking to go to the meeting but couldn't find the club today either," she thinks to herself.

As represented in the vignette, the social environment of a college campus can be intimidating for first-generation college students (FGCS) because they are less likely to possess an inventory of social and cultural capital that their peers enjoy transitioning to the college setting. In this instance, Camila did not intuitively know how to join a student organization if that group did not appear at the involvement fair. Who could be a resource in order to get her questions answered? How would she identify administrative offices that would be helpful in guiding her with the next steps?

With whom and how could she process this frustrating experience of navigating the campus without feeling embarrassed? These were questions she did not have answers for, nor did she know where to go to get them answered. And what about FGCS academic experiences? How do FGCS access campus capital necessary for academic success in college? How do FGCS acquire the campus capital that can aid their academic progress? First-generation college students like Camila can struggle socially and academically. They are more likely to withdraw from college before completing their degrees at rates higher than those of their peers with college-educated parents.

FGCS ponder questions like these in their social transition to campus life as illustrated in prior chapters, and students like Camila are often left stymied when they encounter barriers to navigating the social setting of the campus. Fortunately, the pervasiveness of social media on college campuses has helped to mediate these feelings in FGCS by providing them convenient, immediate access to resources and connections to campus life. Today's college students are consistently connected to social media and the virtual setting of campus life through their mobile and laptop devices. Because of this, college students are easily able to disseminate and receive insight from their peers regarding student life, club programs, and leadership opportunities through posts and messages on Facebook groups, event invitations, and viral marketing campaigns.

Additionally, technology and social media have effectively enabled students to more easily maintain connections with their closest inner circle networks, known as strong ties, while in college through texting, virtual chatting, and video calls. The capabilities enabled by technology and social media effectively ease the transition into the college setting, while simultaneously ensuring students continue to receive the needed support from their family and friends to navigate this uncertain time. In this chapter we follow FGCS students like Camila through their collegiate experience, paying specific attention to how their social interactions are influenced by technology use as students navigate their college environment.

Strong Ties and the College Transition

I try not to let the stress interfere with my grades and everything. . . .
I still have a very good GPA, but that wasn't me at
100%. . . . I could have done better if my family was closer.
It would have been great, but they're far away.
—Ava

Although it is not unusual for college students to feel disconnected from their peers and the social setting of campus during their transition into a college environment, these feelings of uncertainty can be especially intense for FGCS, as Ava described above. FGCS typically arrive at college with less social and cultural capital than their continuing generation college student (CGCS) peers, leading FGCS to have difficulties in perceiving how to most effectively integrate into the campus social setting. This lack of inherent capital during and throughout the transition to college exacerbates the separation from familiar networks that were crucial for supporting students as they navigated the social, cultural, and academic challenges faced in high school. As Krista stated: "I find it really important [to stay connected with family and friends] because I've always been afraid to be distant from them. They're important to me and support me through everything. I can't always be there to see them, so I try to stay connected as much as I can whether it's through online or on the phone." Krista's reflection demonstrates how FGCS require active encouragement and recurrent communication with parents, siblings, mentors, and peers throughout the college experience. Although the physical distance separating FGCS and their families presented logistical barriers for face-to-face (F2F) interactions, virtual connections through mobile devices and online social media provided convenient and effective means to help students maintain and preserve these relationships while at college.

Given FGCS separation from their strong ties and the lack of social/cultural capital to most effectively navigate higher education, Millbrook University invited FGCS who were provisionally admitted to the institution to participate in the EOP as described in the Introduction. These students were the participants in our study, and were mostly racially minoritized, non-White students within a predominantly White, upper-socioeconomic status (SES) campus population. This context initially contributed to feelings of otherness and a decreased sense of belonging for EOP students who did not perceive themselves as fitting in with the general student body, but this social gap was somewhat bridged by their ownership of tablet devices provided by the EOP. As Meena described: "Yes, like I feel like almost every student has [an iPad], so having my iPad gives me a sense of belonging, too. Like now, everyone says you have an iPad too. Let's do such and stuff. You feel really comfortable because a lot of students already have their own iPad, so when you have one, it's kind of like you kind of fit in."

Students also identified student organizations on campus that were instrumental in validating their non-White identities, and easing the social transition by connecting them with culturally similar peer groups. As Nico explained:

> It's been a difficult transition, the first semester definitely. Going to a high school that's predominantly Blacks and Hispanics, it was kind of like a culture shock coming to Millbrook, but I expected that coming in. I just didn't really know how to deal with a predominantly White campus since I wasn't in a similar environment back in high school. But, I joined a club here and went on a service trip to the Dominican Republic. That group has been like a second family to me on campus, and that's kind of been my gateway to meet other people.

These connections to friends and peers were essential in tempering the initial feelings of isolation, foreignness, and anxiety that are typical for FGCS throughout college.

Another way that the FGCS would cope with the unfamiliar, foreign college environment was to turn to their strong ties back home. These online interactions provided positive reinforcement for FGCS by giving them access to cultural capital that reaffirmed their worth and motivation to attend Millbrook. Connecting to their support networks back home was a regular occurrence, but students selected different technologies to connect with particularly strong ties. As proxies for in-person communication FGCS generally preferred to connect with their parents through mobile calls and video chatting. As Ava described, "I want to kind of keep a conversation going throughout the day with family. Usually I phone call at night, or we FaceTime a lot. FaceTime is better with the iPad."

The ability to see and/or hear parents enabled FGCS to be virtually present in both the unfamiliar college environment and the intimate home locations with which they were most familiar. These intimate and personal communications enabled parents to be virtually present in the college setting to support their children in their transition, something relatively impossible given their class and economic contexts. Not insignificant, these online communications also allowed FGCS to feel like they were able to continue to offer support to their families despite their physical absence. Carolina elaborated on the importance and means of communications with her family during an interview with a member of the research team:

CAROLINA: It's important [to connect with parents] to make sure everyone else is fine. So it's nice to know how everyone else is doing, whether they're doing poorly or if they're doing really great. It's just nice to know how everything else is going outside of Millbrook.

INTERVIEWER: So how do you usually communicate with them?

CAROLINA: It would be either just text or call them. Maybe through Skype every now and then, yes, maybe Skype now and then, or just send texts here and there.

INTERVIEWER: So what do you gain from those relationships?

CAROLINA: I guess peace of mind, just knowing that everyone's fine, make sure they're doing well.

Certainly, familism motivates Carolina's online communications. Loyalty to family and a positive regard for familial interdependence characterize her strong bonds and the capital that these bonds provide her. As researchers have noted among Latino/a populations specifically, familism provides support for Latino/a college students, positively impacting identity development and academic performance (Ong, Phinney, & Dennis, 2006; Torres, 2004), and others note that familism serves as a cushion for symptoms of depression and acculturative stress among Latino/a college students (Cheng et al., 2016). Among our students, maintaining strong bonds with family through online communications defied the assumption and conviction common among higher education professionals that students' developmental need for autonomy necessitates "distance" from parents and family (Schiffrin et al., 2014).

The decision whether to make a phone call/video chat, send text messages, or use social media to connect with someone varied depending on the individual whom students interacted with in their communications. Students differentiated how they chose to communicate with someone depending on the relationship. For instance, FGCS rarely, if ever, cited direct communications on Facebook as the preferred method of connection with their parents, but rather used the social media platform to indirectly inform family about their experiences, endeavors, and successes on campus. As Carolina stated, "Facebook is more like just a social place for me. I don't use it to talk to my parents. I use it with friends for quick communication. Facebook is a quick way to get in contact with friends because they're always available online." As Carolina asserted, there was a delineation regarding the most comfortable means with which students communicate with their parents. This distinction of communication modes can be attributed to a number of explanations that include access (many parents do not have Facebook accounts, or laptops/smartphones to access it regularly), but also the fact that students were likely to maintain the norm that social media were most appropriate for communicating with peers, not parents.

Connections with friends from back home, as well as with siblings, also took place through social media. Students would often describe how Facebook or Instagram was a means to maintain relationships with their strong ties by browsing pictures on profile pages and status updates on important life events. Students used social media to post pictures or status updates as a window into their personal experiences at Millbrook. Ella explained, "Facebook is important because my high school friends they go to different schools so we don't always have time to talk. It's always nice to see like what they have been doing like with pictures or status wise." Meena further contributed to this sentiment that social media were an effective means to maintain relationships with strong ties at home:

> Like I said, nobody has time to call everybody on their phone to let them know what's going on, so Facebook is kind of the way to be a mass media as far as putting something up and people kind of keep tabs on you. Like people in my family, if we meet up at a gathering at something, somebody will say oh yes I saw that you did such and such, or I saw your pictures, you guys look so nice, or people will comment their little comments and say stuff like so proud of you and things of that nature. So that's how we pretty much stay connected.

An additional benefit of communicating with friends back home during their college transition through social media was that such interactions did not mandate continuous time commitments often required by phone calls or video chats with parents. Students could initiate conversations with peers or siblings that took place over the course of multiple days through social media messenger apps, permitting FGCS to access cultural capital by simultaneously interacting with a number of strong ties at any time and place during the day. Krista resonated with the convenience of social media messenger apps: "I think Facebook Messenger is the easiest way to keep in touch with my friends back home because everybody is constantly checking the site and sending messages. It's so easy to just open up your laptop and send messages whenever you want because the site is always there." These less structured connections were frequent and did not require an uninterrupted time commitment mandated by phone calls. Students agreed that the flexibility provided by Facebook Messenger and Snapchat was most appropriately catered to communicating with strong ties across similar age groups, but not specifically designed to connect with parents and older family members such as uncles or aunts. Regardless of how students interacted with their strong ties, it became clear that these virtual connections and interactions with their networks back home were critical in easing the transition of FGCS into their lives as full-time college students. Ava clearly articulated the importance of connecting with family during her first year at Millbrook:

> Yes, me being homesick, I know it's just I'm a freshman and all that, but like I said I'm a very family, community-oriented person, and I never realized how much my community meant to me and was a part of my life before I came here to Millbrook. What kept me going in high school is just me knowing I had this responsibility to my family so I had to go to college. Sometimes I'm so far from my family now, I don't feel any motivation to just keep going, but I have to think about the future and the long run when I talk with them. It's just the love that I feel from my family that motivates me to keep going.

Given the difficulties that FGCS are likely to encounter in their transition to college, technology and social media provided a means to mediate the barriers that often inhibited effective integration in the college environment. This is due to the ability of FGCS to easily maintain connections to their strong ties using social media as a means to establish their presence in both the familiar settings of home and the new environments of a college campus. The experiences of students in this study reinforce the fact that cultural capital provided by strong ties is crucial to the success and foundation of support for FGCS on campus. Strong ties were essential in instilling the qualities of perseverance and resilience that fortify the strength required for these FGCS to succeed and persist in college. These connections provided a foundation of support and faith, enabling students to rely on the enduring presence of strong ties as a means to reinforce the resilience that was required to navigate a culturally unfamiliar space.

Weak Ties and Social Media in the Out-of-Class Experience

> I use Facebook on campus. I mostly am connected with some student leaders who are involved with other organizations. They usually post certain things about various groups or a club I'm interested in. That's how I get information about clubs and that's how I would contact someone when you're not close enough to know that person as a friend. Facebook is good for that kind communication.
>
> —Nico

Although strong ties and social media played an important role in helping FGCS persevere through the transition into college, peer relationships in the form of weak-tie relationships and social media also proved instrumental in advancing the social life of campus. As previously noted, weak ties are the connections students have with individuals who are not most like themselves, are less intense relationships, and are more plentiful. These connections include classmates, peers in student organizations, roommates, and campus mentors who are sources of bridging capital that assist FGCS in their social transition in college. An example of how weak ties were beneficial to our students is represented by Gabriel, who reflected on how those in the EOP enabled him to become involved on campus through an involvement fair: "I found out about the activities fair through the [EOP] mentors who are part of a few of the student groups on campus. I met and talked to the mentors since we have several weeks together in the summer. As you talk about your interests, you create bonds and relationships with them. Knowing someone on campus was helpful because it got us to open our eyes to other different student groups on campus." This example of weak-tie relationships playing an active role in facilitating bridging capital is a necessary interaction to ensure FGCS are exposed to campus resources and opportunities for social integration. The EOP was instrumental in fostering weak-tie connections that supplemented FGCS lack of campus capital, and further primed students with interactions that could reduce the stresses of transitioning into the social life of a university setting.

FGCS relationships and interactions with weak ties were not exclusively restricted to F2F interactions. These communications and the facilitation of campus capital were also reliant on the integration of social media such as Facebook into everyday life. In an instance that typifies the instrumental role that social media play in facilitating campus capital between weak ties and FGCS, Meena described the process of getting involved in a student club: "So actually, I have a friend that I met through joining a culture show together, and she told me how there was a different culture show for the Hawaiian club. She asked me if I was interested and asked me if I wanted to participate in it. I was like, 'Yes, sure. I think I would be interested.' She basically said, okay, I'll send you the Facebook event link and just sign up through there and then she did that." This interaction exemplifies the significance that a student's numerous weak ties and social media possess in effectively facilitating campus involvement for FGCS at Millbrook. Students agreed that traditional flyers and posters are no longer the primary means for disseminating information about campus involvement and club programs. Rather, online networks and social media such as Facebook more effectively facilitated social involvement on campus. Gabriel confirmed this sentiment: "If I find an event that I think is interesting I will be like, hey, I will try it out. I mostly hear about it through Facebook. I know they have flyers all over the place, but I rarely look at those." These testimonials confirm social media are the primary and preferred locations that students rely on to become integrated within the social life of a college campuses. However, FGCS Facebook feeds are curated and most often are sourced through and by known weak ties rather than new, unfamiliar weak ties. Historically, college students have used Facebook not to make new weak bonds but rather to keep track of already established weak bonds on campus (Martínez Alemán & Wartman, 2009). Like Meena, among our FGCS, capturing campus capital through Facebook was more often a consequence of extending homophilous weak ties or widening homophily in online social networks.

The experiences of students in this study further confirm that Facebook occupies the designation as the virtual hub of campus life where clubs, programs, and social opportunities are circulated on a consistent basis. When asked where she would go in order to find out what events or special announcements were happening on campus, Yvette responded, "Usually online, like through Facebook. Everyone posts in the Class of 2016 Facebook group. You can find anything there, like all of the athletic events are always posted up there. And people looking for roommates. The [student government] and campus opinion polls are always posted up there, too." Students largely agreed with Yvette's reflection on the importance and visibility of Facebook as an integral aspect of campus life. Others went further and explained the nature of Millbrook's campus life being reliant on social media specifically necessitated Facebook be used as an everyday tool in order to effectively feel like they were incorporated into student life on campus. Mason, echoing the sentiment of others, explained how he did not enjoy using Facebook, but simultaneously admitted that it would be difficult to live without the social media because of how dependent he was on the site at Millbrook: "I started getting bored with Facebook, so I initially got rid of it. But without it, it's just so hard to keep in touch and know what's going on. That's honestly the reason why I have to keep Facebook now." Mason's feeling was echoed by Carolina, who maintained, "I've thought a lot about deactivating my Facebook, but then I think again. Like, how am I going to find out about the events on campus? How am I going to get people to know about my club events? When I think about not being able to do that, it would suck." The inability to effectively stay connected with campus life at Millbrook without Facebook left students with the understanding that engagement on social media would have to become a natural and everyday aspect of student life. Since its early presence on college campuses, Facebook has been a kiosk for campus information (Martínez Alemán & Wartman, 2009) and for the most part, our FGCS accepted this reality and more intentionally engaged with the social media site to maximize Facebook's potential for social involvement and visibility on campus.

Students described Facebook as a platform that was largely purposed for the promotion of Millbrook's extracurricular life in a number of diverse ways. Students at Millbrook consistently used Facebook to create event invites for club programs, to post virtual flyers to advertise organizational gatherings, and to change user profile pictures as a way to publicize upcoming programs and club meetings. Krista provided insight into the student culture of online Facebook marketing by describing her actions on the social media site: "Some of the clubs that I do on campus require me to post stuff on Facebook. I've been posting so much stuff that is not about me at all. One of the groups I'm in on campus has a show coming up, so I'm constantly being told 'Make our event your status! Change your picture to our flyer! Send out messages about the date of the event!' And so, that's how I mostly use Facebook." Students acknowledged that these online strategies were largely effective in drawing attention to club events online. They also recognized that the confluence of profile photo updates, event invites, and constantly changing newsfeeds on their Facebook accounts was a convenient location for accessing bridging capital through the multitude of opportunities for involvement at Millbrook. In turn, Facebook was categorized as a rich and convenient forum that had the capabilities of connecting students to multiple facets of campus life.

Facebook was also referenced as a critical tool that was effective in developing, facilitating, and strengthening weak ties on campus. Students would often leverage Facebook's ability to be an informational directory of their peers to more strongly develop relationships on campus. Gabriel elaborated on his experience of advancing friendships with peers on campus through Facebook: "I kind of use Facebook to develop friendships, too. I was a part of this co-ed basketball team, and at first we had played together, they requested me on Facebook so they could see the person I was since we didn't know each other face to face. We'd just play basketball then, say 'See you next game' and leave. . . . Facebook was the way we got to know each other better." This reflection by Gabriel further emphasizes the reliance that students have on social media at Millbrook as a forum not only for acquisition and transmission of capital, but also as a tool that effectively initiates, bridges, and reinforces peer connections in college. These relationships then become fruitful connections that help further indoctrinate FGCS into campus life. This process is best exemplified by Arun, who elaborated on how a connection with a peer on campus was instrumental in helping him get involved through social media: "I live with two people that were in [the EOP] so, like, usually one of my roommates always forwards me stuff about a club on Facebook if he thinks I might also be interested in it. That helps me out a lot with finding opportunities on campus I wouldn't have known about myself." These declarations of the importance of Facebook, and the weak-tie relationships it fosters, for FGCS symbolize the essential cornerstone of student involvement played by social media at Millbrook. These findings further confirm that Facebook served multiple purposes for FGCS in this study that included providing convenient access to the virtual setting of student life, facilitating connections with weak ties, and circulating social capital that aided in their social integration. These outcomes have all been found to effectively enhance the higher education experience for FGCS, and were identified by our students over the course of multiple years throughout their undergraduate experiences.

As students developed connections with weak ties on campus, the means by which FGCS maintained and strengthened those relationships was often through private messaging. These communications were conducted primarily through individual text messages, correspondence on Facebook Messenger, or temporary messages on the mobile app Snapchat. These individual communications reflected the prolific and regular nature of online interactions at Millbrook but also comprised the back channels to campus life. As back channels, these communications were "private" online communications and their consumption was purposely restricted.

The Messenger application within Facebook, Snapchat, and the mobile text-messaging app GroupMe were three commonly cited tools for coordinating a group of peers. These applications permitted students to simultaneously communicate with multiple peers through the convenience of a mobile device, an iPad, or a laptop without requiring a phone number. Nico explained in an interview how he communicated with his friends on campus:

NICO: GroupMe is great. You don't even have to have the person's number. If they have the app, you just find their name and you can message them. It's an app through your phone. You can download it on iPad, too. It's pretty awesome. I don't know anyone who doesn't have it.

INTERVIEWER: Really?

NICO: Yes, because you don't even need to know the person. I mean you need to know . . . like if you have their name, that's all you need.

The short-term interactions on messaging apps almost exclusively comprised communications with weak ties on campus and these interactions were conducive for FGCS to employ because weak ties were often identified as temporary and plentiful in number. Due to these characterizations, students sought technological ways to interact with weak ties that were convenient and temporary in nature. Meena described why she communicated with peers on campus through the ephemeral messaging app Snapchat: "I think Snapchat is easier to use here at Millbrook. Like, it takes five seconds and you don't worry about it. You don't think about what you sent a minute after. It's just there when you need it and it's then you're done." Camila regularly uses Facebook Messenger or text messages to interact with peers and affirmatively stated, "No, I don't call anyone" even if last-minute changes or cancellations of plans needed to be made.

The experiences of our students confirm the importance of forming robust relationships with weak ties in conjunction with establishing a regular presence on social media. Although student connections to strong ties from home were vital in helping FGCS maintain persistence and grit during their collegiate journey, weak ties on campus were fundamental in facilitating community connections and social involvement through in-person and online interactions. The bridging capital gained through weak ties helped promote familiarity in the foreign space of a college campus that led to increased involvement and engagement. Additionally, social media, not emails, were effective in helping connect to students and temper the overwhelming cultural environment of Millbrook. Social media also provided a forum where the acquiring of campus capital was both streamlined and simplified, allowing FGCS to develop a foundation of multifaceted peers and knowledge that elevated positive experiences throughout their college experience.

Threats to Circulating Campus Capital on Social Media

INTERVIEWER: Do you feel like you are part of the community here?

KRISTA: No. Well, of course like I'll probably feel at home at the Black Student Forum, or if I went to like an [Office of Multicultural Affairs, or OMA] event. Campus is sometimes hard to relate to.

FGCS ability to effectively leverage connections to strong ties, weak ties, and social media for the facilitation of campus capital and to integrate within the social culture of Millbrook were threatened at multiple points by both peers and administrators. Students in this study raised a number of concerns regarding the psychosocial impact of encountering the uncomfortable and adversarial peer culture of the predominantly White student body at Millbrook throughout their time on campus. These behavioral and racial strains transcended the physical campus setting and migrated onto social media platforms, impeding students' engagement online. Students often cited the inability of campus administrators to comprehend the basic capabilities of social media and the lack of implementing online strategies to connect FGCS with campus resources, programs, and departmental engagement with the student body as issues that required remediation. These specific cultural and administrative barriers to the acquisition of campus capital and social integration are discussed in the following section.

PEER CULTURE AND RACIAL CLIMATE

During their initial transition to Millbrook, FGCS in this study often described the campus environment and social life as unfamiliar and incongruous with their home contexts. Social interactions during their transition were not often well received by FGCS, as the unfamiliar environment of socializing with predominantly White peers situated in a "drinking and partying" culture was a source of conflict for students unfamiliar and uncomfortable with this peer culture. In describing the campus environment, Meena reflected: "I went to a high school where it's so diverse, and coming here to Millbrook it is not as diverse as I had expected it to be. The culture here is a lot different. The drinking culture and partying is a lot crazier than I thought it would be, and I don't drink. I just haven't found a group of people that are similar to me yet." Meena's reflection resonated with students in this study. Students often cited discomfort fitting in to the Millbrook peer culture and maintained distance from the pervasive drinking and partying that occurred on the weekends. These incongruencies between FGCS expectations of social life and the Millbrook peer culture of excessive drinking and partying persisted throughout much of the students' experiences at Millbrook and were a perennial source of frustration for students who wanted to more closely identify with the campus community.

Students also were cognizant of how their identities as racially minoritized students at a predominantly White campus contributed to difficulties feeling fully integrated with the social setting at Millbrook. When asked to describe their sense of belonging and if they considered themselves as part of the Millbrook community, students would often reply with observations such as Nico's: "Socially-wise, I feel like there's a culture divide on campus. I feel like there's not a racial mixing on campus and stuff. Like, the racial separation is still very real." In other cases, students such as Arun were adamant about their assessment of their sense of belonging:

"No, I don't feel like I'm a part of the community here." Racial segregation on campus and FGCS self-described status as nonmembers or outsiders to campus culture lessened their desire to foster relationships with peers who could potentially serve as sources of campus capital. The aversion to connecting with White peers was fully captured by FGCS documentation of the regularity and pedestrian ways that racism manifested itself on campus.

In one disturbing instance of racial insensitivity by her White peers, Carolina described her feelings toward a racially themed party that was held on campus by one of Millbrook's prominent athletic teams:

> So basically last year the team had a party, and the theme was CMT, so Country Music Television versus BET, Black Entertainment Television. That's so ignorant to me. It's just so ignorant to me that no one on that team thought oh . . . because I feel like there's a stereotype to Country Music Television, like cowboys and things like that, like you wear the cowboy hat, cowboy boots. That's something that you can, I don't know, imitate. But BET, what does BET mean? What does dressing up as a BET star mean? So it's kind of like I have no idea who came up with that theme, and then also the fact that no one thought this is a bad idea, like, guys, maybe we shouldn't do this. Like no one said that. Everyone agreed with it.

Almost all the students in this study described their frustration with the inability of the predominantly White student body to recognize the racial tensions on campus that erected a barrier to FGCS full membership in the Millbrook community. The presence of this hostility proved detrimental to facilitating positive perceptions toward the campus's social life, and often daunted students in their desires to more fully integrate into Millbrook.

In response to this adverse campus racial climate, students often sought out ethnic subcommunities, or maintained close friendships with more intimate circles of friends such as roommates or other summer bridge students of color (SOC). These SOC peers validated students' experiences as racial and ethnic minorities on campus. For example, while Krista did not identify with the predominantly White student body at Millbrook, she found community by joining the Caribbean Culture Club and Black Student Forum student organizations. The experience of being an active member in student groups with culturally familiar peers was instrumental to enabling increased campus connections with like-minded peers, which in turn advanced both a sense of belonging and accrual of campus capital through these trusted weak ties. When asked to reflect on her time in the Caribbean Culture Club and the Black Student Forum, Krista stated, "I mean, I attend organization rallies and stuff whenever they have something on campus. So whenever there's something going on, I always tend to know about it. Friends are always constantly reaching out to me asking me if I'm going to this or if I'm going to that. I do feel very present, so yes, I do feel like I belong."

Krista's reflection on her experience in these cultural clubs describes the possibilities of both campus belonging and the circulation of campus capital when students feel fully valued and comforted by their weak ties. In this instance, Krista's engagement and capital were scaffolded by the ethnic subcommunities of Millbrook. Conversely, she experienced less robust connections and accrued less capital from the dominant partying, drinking, and White peer culture of the larger campus community. The contrasts between these campus contexts draw attention to the cultural divides that hinder and impede the facilitation of campus capital and social integration at Millbrook.

FGCS repeatedly cited on-going racial tensions as a problematic issue with their transition and integration on campus. Racial tensions on campus (racial separation, lack of non-White students on campus, racially themed parties) extended to online spaces. FGCS were also aware that social media were a context in which racism, prejudice, and discrimination were widespread. Specifically, the emergence of a geo-specific, user-anonymous social media app, Yik Yak, which ceased to exist in April 2017, proved a controversial presence on campus. Students often described encountering anonymous statements that encouraged violence against the Black community, posts that used racial slurs, and other degrading comments about Asian and Latino/a communities. Although the racially insensitive posts on Yik Yak were anonymous, FGCS hypothesized these offenses originated from White students based on their assessments of the campus racial climate.

FGCS described anger, annoyance, and fear in response to these online hostilities. As expressed by Camila, "It's just frustrating that people say all of these things. I get frustrated when comments become hateful. I do of course get angry." In another instance, Andrew raised concerns about the well-being of non-White students on campus after violent statements about harming Black students emerged on Millbrook's Yik Yak feed: "Like just because Millbrook is safe doesn't mean that people can't get killed easily."

Students often cited the prevalence of racialized offenses on social media as a reflection of a hostile environment that posed another barrier to their effective college transition and success. The presence of these communications on social media further typified the racial tensions that existed on campus. The frequent hostilities on Yik Yak made FGCS suspicious and wary of White peers. Additionally, these online aggressions suggested students of color (SOC) were not welcome in particular social media platforms (Yik Yak and other anonymous online forums).

To counter these online racialized offenses, students heavily curated their peer connections on their most used social media (Facebook, Instagram), and avoided apps such as Yik Yak where racial hostility was likely to proliferate. Students also advocated for interventions to eliminate the perpetuation of racism on social media and proactively engaged in advancing antiracist dialogues online through the hashtag #BlackLivesMatter. Although students themselves actively engaged in these online practices, they did not make it clear if and how campus administrators should also be involved in these virtual dialogues. What is clear, though, is that all aspects of social media, whether positive or negative, were fully integrated as part of campus life that affected the experiences of FGCS.

The outcomes of encountering online racial hostility are concerning for a multitude of reasons. Beyond the ethical care of students, psychological distress caused by encountering racism, and the possibility of danger to the physical well-being of students at Millbrook, the presence of a hostile online environment effectively tempers FGCS use of social media in multiple ways. First, students were likely to curate connections in their user networks to avoid peers who were likely to perpetrate online hostilities. Such an act serves to limit the connections that FGCS made with weak ties on campus to more homophilous individuals, effectively reducing the ability to gain bridging capital from weak ties, due to an increased need for cultural capital from like-minded and culturally familiar peers.

Second, students became more averse to using certain social media platforms that are likely to promote racial hostilities. This is troublesome because testimonials from FGCS assert that social media, regardless of the functionality, are effective tools to enhance their social experiences. When the presence of racial hostility discourages students from using social media, it is less likely that campuses will fully realize the potential to leverage the positive capabilities of social media for engaging all students online for the purposes of circulating campus capital and promoting social involvement. Instead, online racial hostility limits the scope of FGCS enthusiasm for engaging across the spectrum of social media on college campuses, thus limiting the potential for enhanced connections with weak ties and the circulation of campus capital.

The considerable and real racism cited by students on social media and its noticeable manifestations in campus life is a further representation of the sizable influence that social media have on today's college. Such a ubiquitous threat to the social integration and the facilitation of campus capital requires attention and resolution by campus administrators and educators working with FGCS.

It seems, then, that FGCS used their iPads, mobile devices, and laptops to maintain strong bonds with family and home networks and in doing so, secured bonding capital that appears valuable to their transition and college persistence. FGCS used social media primarily to enrich weak bonds and access bridging capital with homophilous peers, to access campus capital through information feeds, and to connect selectively with mentors.

Through their varied relational ties, FGCS obtained campus capital in different ways. Through the strong ties with people most like them and who were familiar with their experiential upbringing and circumstances (family, friends, and teachers from home, and maybe even other FGCS students), FGCS sought out emotional support or bonding capital that could buoy their confidence and lessen their insecurities about belonging at college. It was the weak ties in their social network (CGCS, faculty and staff unlike them who have insider campus capital), however, that contained the instrumental campus capital or bridging capital that could provide them with valued information for succeeding and belonging at college.

This chapter's evidence of both the positive and negative effects of technology and social media on FGCS experiences provides insight into the experiences of today's college students for campus administrators. The intention of the documentation of these experiences and their impacts on student life is to promote the development of both interventions and policy that best advance positive social experiences for FGCS on college campuses.

5

Bridges to Campus Capital in the Classroom

WITH CONTRIBUTIONS BY JONATHAN LEWIS & SARAH KNIGHT

Lilly has been teaching in the Educational Opportunity Program (EOP) for almost six years and finds it to be both professionally challenging and rewarding. As a first-generation college student (FGCS) she relishes the opportunity to support the EOP students and witness their excitement about being in college. She takes special pleasure in sharing her favorite pieces of English literature, knowing that for some of the students, this is the first they've been exposed to specific works. Over the years, Lilly has come to see herself as a mentor for students who complete the EOP and values the relationships she has with them. At the same time, she feels responsible for teaching academic content and skills that will prepare them for the rigors that await them in the general curriculum.

When the director of the EOP introduced the iPad initiative three weeks prior to summer classes, Lilly was initially excited. She was eager to experiment with integrating technology into the classroom, which she observed other Millbrook professors doing. However, that excitement quickly turned to frustration when it came time to plan how to use the iPad. Beyond a little Facebook, Lilly barely used technology in her own life. She had an iPhone and used a few apps to read the news, find local attractions when she traveled, and listen to music. Yet, when it came to classroom usage, she felt inexperienced and unsure how to move forward. Upon receiving the iPad and an accompanying list of possible apps from the director of the EOP, Lilly tried to imagine using it as part of her teaching, but quickly realized she did not have the tools to use it in instructionally useful ways. With only a brief training in how to use the iPad and only three weeks to prepare for what she imagined to be a major shift in her teaching, Lilly started to feel resentful. She found the iPad to be in conflict with her otherwise laid-back style of teaching. In many ways, it felt forced rather than chosen as a tool to supplement her instruction.

After consulting with the other EOP English instructors, Lilly adopted the use of two apps. She used a poetry app, which was being used by the other instructors to expose students to a diverse group of authors. She also encouraged students to use the dictionary app when they came across terms with which they were unfamiliar. She felt that both apps helped her students access the content she was teaching; however, she never felt fully satisfied that she was effectively leveraging technology to support their learning in a meaningful way.

Campus Capital in the Classroom

Although technology can improve and enhance the ways in which students obtain bridging capital, this process does not occur automatically. Because weak ties are not as robust and, with few exceptions, are unlikely to grow into strong ties, information that is transmitted through these connections must be purposefully structured and carefully deployed in order to serve as an effective bridge to academic knowledge. Faculty members are frequently the lynchpins in this process. Faculty who make effective use of technology within the course environment help first-generation college students enhance their campus capital, which, in turn, can positively influence their sense of belonging. As we can see from Lilly's experience above, this process is not easy and can often be frustrating during implementation.

Faculty who utilize technology effectively to help students attain bridging capital employ varying levels of structured engagement in and out of class, including electronic books, social media, apps for research and note-taking, and virtual office hours, among others. Students view faculty who forbid students to engage with technology, or fail to use commonplace course enhancements like PowerPoint slides, as anachronistic. Simultaneously, many understand that faculty who restrict technology use are aiming to avoid in-class distractions.

Outside the curriculum, students acquire bridging capital from professional and student staff primarily in person, although evidence exists that peer networks—student leaders, peer mentors, and classmates—also readily transmit academic knowledge through mobile technologies, including apps and social media. Without much direction, students will use technology to attain bridging capital from their peers; however, when engaging with the formal curriculum and co-curriculum they follow the lead of faculty and staff, who vary widely in their usage. Intentionality may be the key difference in determining which specific weak ties are likely to use technology to help students acquire the bridging capital they most need to be successful.

Faculty Interactions and Bridging Capital

Although faculty may rightly assume that most students possess familiarity with Web 2.0 technologies, students' knowledge of these technologies does not necessarily transcend the social realm. Knowing what applications, websites, and other resources would best support their academic success is not a given, and without guidance or direction few students are likely to find and utilize these tools independently. Instead, students typically learn about both general and content-specific academic technologies and resources from faculty and staff. Once students have identified the technological tools best suited to enhance their studies, however, they can and do use them effectively to gain bridging capital and support their learning.

In addition to being aware of the academic applications of Web 2.0 technologies, students must be permitted to use them. As seen in the case of Arun's sociology class in Chapter 3, bans on technology due to concerns about student misuse in class, though understandable and not uncommon, can inhibit students' ability to engage with the material and, subsequently, their sense of belonging at the institution. Furthermore, although allowing students to use their preferred devices in class supports their learning, instructors' creative and active incorporation of technology in the course content promotes an even more enriching learning experience. Faculty can also support students' growth and learning through technology outside the classroom through strategies such as virtual office hours.

USING AND SUPPORTING TECHNOLOGY IN CLASS

I think with my generation we are, we get bored very easily, so technology is engaging, especially because that's always what we're surrounded with. So I just feel like it's, I feel like when, because we're always using this technology and we're always in this zone, if we have to go and listen to some person talk for a while it can just get very boring very quickly, even if the subject matter is interesting we can make an excuse for it to be boring. So I just feel that technology has potential. Like the PowerPoints, they're captivating. And also, it's good to see the words on there. It's much easier to take notes.

—Mason

As Mason describes, today's college students are almost constantly immersed in some form of technology, which they view as contributing to a more engaging method of instruction in the classroom. Consequently, the absence of technology can significantly affect the students' attention to the lesson, even when they find the course material interesting. In contrast, incorporating even simple media or technology, such as PowerPoint presentations, better facilitates their class participation.

When asked about whether or not their instructors supported or used technology in the classroom, our FGCS reported a continuum of integration ranging from nonexistent to high. On the lowest end of the spectrum, faculty neither permitted nor incorporated technology in the class. Some instructors were explicit about why they forbid the use of electronic devices during class—to prevent misuse and distractions—but even when instructors did not explain, students perceived the reason to be the potential for distraction and were understanding of this rationale. Meena stated: "I completely understand because I will constantly see kids, if they're in a class with a computer, they're on Facebook while the professor is lecturing, and it's distracting to other kids around them, and it's distracting them, too. So I understand not allowing."

Although a few students preferred to take notes by hand, either to avoid potential distractions or because they felt less technologically savvy, most of our FGCS appreciated when faculty allowed them to use tablets or laptops to take notes or record the lecture. Such permission occurred frequently, even when the instructor did not actively employ technology. As we saw with Arun, students reported that it was easier to take notes with a laptop or tablet, especially when instructors did not use PowerPoints or other visual aids and when they lectured quickly. Not having access to these electronic note-taking tools inhibited the ability to take notes effectively. Referencing a course in which he was not able to take notes on his computer, Andrew complained: "I get specifically annoyed with my religion teacher because she says so much stuff so fast, and we have to take notes, and it's just like I can't even read my notes a lot of times at the end. If I just had a laptop, it would be so much easier."

Additionally, students highlighted the advantage of using their devices to clarify concepts during class: "If you don't understand something, you can always research something as the professor is saying something, so it actually makes more sense," Nico shared. This immediate access to online search engines or other resources, therefore, serves as a source of bridging capital for students to define unknown words and provide alternate explanations or additional information about course topics, thereby enhancing their engagement with and understanding of the material being presented in class.

Students' reactions to the faculty who refused to incorporate technology into lessons ranged from slightly amused and incredulous to irritated. Yvette and Nico described a professor who neither used nor allowed students to use technology in class:

YVETTE: He's very old school. He doesn't even use his own computer.
NICO: He uses a chalkboard.
YVETTE: He uses a chalkboard and then he uses like the old projector thing.
NICO: He is so old school.

Although students understood that some disciplines or classes, such as discussion-based courses, are less conducive to faculty's using technology, they also expressed frustration when faculty refused to employ it in lectures. Camila, for example, was frank about her annoyance with a professor who did not even use PowerPoint slides: "My politics teacher is ridiculous. He just comes in class and talks. He doesn't have any notes at all." If Mason's perspective holds true for the majority of his peers, lecturers like this who neglect to supplement their orations with visual references or those who employ outdated technology ironically run the risk of losing their students' attention and impeding their ability to engage with the material.

Students hypothesized that some faculty's lack of familiarity with technology explained its absence in the classroom, and instructors occasionally admitted that this perception was accurate. Even when faculty feel comfortable with their personal or general use of technology, the ability to effectively incorporate it into their pedagogical practice does not naturally ensue. For example, Lilly, one of the EOP faculty, shared her experience using an iPad for instructional purposes for the first time: "My teaching style is really laid back and casual. And I felt like every time I used the iPad I became a different teaching persona because it was not something that I just picked up and felt comfortable with." Similar sentiments rang true for other EOP instructors, even those who described themselves as technologically savvy. Although they recognized the potential for technology to supplement their teaching practice and support their students' learning, they emphasized the necessity of having sufficient time and training to identify meaningful rather than gratuitous strategies for doing so.

Nonetheless, with access to and basic training on the use of different devices, software, and applications and the time to explore possible discipline-specific strategies, faculty identify and implement technology in a variety of general and subject-focused ways. Martha, an EOP faculty member, commented that being provided with an iPad for instructional use prompted her to "think about some of my teaching in more creative ways than I would have in the past" and to try to make the most of the technology that had been made available to her. Although some tools or methods simply facilitated easier access to course materials and information, the most creative approaches integrated technology in a way that genuinely complemented the content to effectively engage students and enhance their learning.

FGCS indicated that many of their instructors offered or supplied electronic course materials, such as e-book versions of the textbook, or uploaded PDF articles in the course management software (e.g., Blackboard or Canvas). These electronic course material options were viewed as preferable to textbooks because they are more portable, requiring only one device to access multiple texts or documents, and they are frequently less expensive than textbooks. Also, the software programs used to access these materials typically include the same annotative functionalities that paper versions allow, such as highlighting or adding notes in the margins. As a faculty-driven decision, therefore, selection of the format for course materials provides instructors with an obvious opportunity to incorporate technology.

By far the most commonly reported in-classroom technologies were PowerPoint and video presentations. Nico summarized what many of his peers revealed about these applications of technology in class: "I think the most popular way is PowerPoint. . . . A lot of teachers use PowerPoint, maybe movies, sometimes like clips, but I think that's pretty much it. Like just to show us a visual thing. Technology is usually an effective way to present that." Our FGCS appreciated having access to the PowerPoint slides to follow along during lectures. Rather than furiously scribbling notes, they could focus on listening to the instructor and jot down or type additional comments or points of clarification not included on the slides. Video clips enhanced class sessions both visually and aurally, most often with the aim of providing examples or explanations outside text materials. One student, Andrew, described how his professors extended the course material through videos as a way to "give us a different perspective from someone else, not so much the professor and not so much the textbook, but someone else that's in the field to just further the understanding or really make the concept or theory concrete." Several of the EOP faculty and staff also noted how students were drawn to videos and acknowledged the importance of trying to incorporate video media as a means of engaging them.

Less commonly reported general practices for using technology in the classroom included recording and posting lectures, interactive webcasts, and iClickers. When they were employed, FGCS generally found the use of these technologies to facilitate their learning. When a lecture was recorded, students were able to easily review it and check the accuracy of their notes and understanding of the material as they studied. Since few instructors actually record their lectures, many FGCS reported using their own devices, including phone or iPad recording apps, to achieve the same purpose. Similarly, some students used the cameras on their phones or iPads to take pictures of diagrams or other information written on chalkboards or whiteboards. Thus, students viewed both lecture recording and taking pictures as tools to ensure accuracy and efficiency in taking notes as well as digesting and reviewing course material after class. Many of the students seemed to have learned to use these strategies during the EOP by observing the program staff or their peers employing these techniques.

Webcasts or webinars were generally used only as backup options for lectures or review sessions in the event of inclement weather, but students reported being surprised by their usefulness. Initially skeptical of his professor's plan to hold a webinar on a snow day, Gabriel described the session as beneficial: "Everyone got to ask questions. He had gone over the slides, and then we saw each person's response. So if we read it, we had more of an understanding of the other ways it could be analyzed." Although students viewed iClickers for attendance purposes as silly, they recognized their value for attentiveness and class participation, especially in large lectures (most commonly in biology, chemistry, and psychology). As Ava explained, "I feel like having the clicker, and that grading you in participation and quizzes and stuff, it [keeps] you more focused." Mason also described how his organic chemistry professor's use of the iClickers made for a more engaging class: "So he's like, this is a question, you have this amount of time to answer this question, and then you can see how many people choose what. So it makes it more interesting that way because you never want to be wrong."

In addition to the recording and camera apps, faculty occasionally demonstrated and encouraged FGCS to use other mobile apps for academic purposes. EOP staff introduced students to Evernote, a note-taking and organizational app with multimedia capabilities, and more than 30% of the students reported using it at least once per week after the academic year began (Table 5.1). Depending on the specific classes they were taking in the fall, students also continued using academically oriented apps that their EOP instructors had encouraged, such as the Merriam-Webster dictionary app and the literature analysis and grammar apps.

Students and faculty alike indicated the utility of such apps in and out of the classroom. In reference to the dictionary app, Gabriel stated, "In philosophy there's a lot of words that I did not know, so I would go to the dictionary and sort it out and see how that would connect to what we're learning in class." Lilly, an EOP instructor, also highlighted the dictionary app as a helpful tool: "It's great for students just to be able to use that in class without carrying around a gigantic book or something, so that's really useful." As with having access to search engines to clarify concepts during class, the convenience of such apps allowed students to easily and immediately fill gaps in knowledge and increase their understanding of course material. Gabriel's experience provides evidence for the type of classroom experience where first-generation college

TABLE 5.1. Percentage of students who reported using apps on their iPad during the EOP (n = 66–68)

	NEVER	ONCE OVER SUMMER	FEW TIMES OVER SUMMER	ONCE PER WEEK	FEW TIMES PER WEEK	DAILY	TWO OR MORE TIMES DAILY
Dictionary	27.3	13.6	28.8	9.1	9.1	3.0	7.6
iStudiez Lite	85.3	7.4	2.9	1.5	1.5	—	—
Poetry	27.9	14.7	38.2	4.4	10.3	1.5	1.5
Adobe Reader	35.8	7.5	19.4	11.9	20.9	3.0	—
Literary Guide Analysis	45.6	5.9	23.5	5.9	17.6	—	—
Grammar	50.0	11.8	17.6	5.9	11.8	—	1.5
Evernote	39.7	7.4	17.6	7.4	17.6	4.4	4.4

students can use technology to attain vital bridging capital and maintain or enhance their sense of belonging in the academic environment. Importantly, the EOP peer mentors observed that the students were more likely to use the apps that their professors introduced and supported as related to course material:

COLIN: A lot of them would either read things on their iPads for English or . . .

RUBY: Especially No Fear Shakespeare.

COLIN: So then they would have that up, and then their paper and then the book. And you could tell they were doing their homework for that.

INTERVIEWER: So they found No Fear Shakespeare on their own?

RUBY: They were, I think they were encouraged to use it by their professors. Provided that they reference back to the original text when citing it in their paper, or something like that.

Thus, by identifying No Fear Shakespeare as an approved supplemental learning tool, the instructors not only facilitated their students' comprehension of the material, but also established expectations for appropriate usage of external aids. The EOP faculty also noted that students seemed more inclined to follow their lead with regard to using new technologies. Martha reflected about using her iPad in class: "I think students are more likely to use it if I am using it and setting an example. Because I'm using it and I'm working with it, so they're all kind of having this experience."

In contrast, outside the context of the EOP, identification of apps to supplement academic content was infrequent, but students did mention one or two instructors who incorporated Twitter into projects or assignments. A few others were reported to have suggested and encouraged their students to use discipline-specific apps. For example, Meena's French professor recommended that her students use Yabla, a language immersion app, and designed assignments that required students to use the app. As with the general uses of technology in teaching, however, the challenge for faculty of incorporating apps into course content lies in knowing about different options and how they can be used most effectively. Given the limited frequency with which students mentioned apps that were not introduced during the EOP, it might, therefore, be inferred that most faculty are unaware of appropriate apps and/or lack the time to search for them.

In addition to using technological devices to present or supplement course material, faculty can creatively incorporate the technologies with which students are familiar into the classroom. Lilly, for example, designed an assignment for a unit on *The Merchant of Venice* that required students to create Facebook profiles for the play's characters in order to bring them alive in their reading. She described the effectiveness of contextualizing the script:

> We wanted them to really understand when you're acting you're not just reading lines. You need to embody this performance. So for me I feel like Facebook and Twitter and Instagram are the places that they embody who they are and what they think is going on. So I really loved, and this year—we've been teaching the same Shakespeare play for a few years—this year my students killed it. They were so excited, and it was amazing. I think a lot of it had to do with seeing Portia or Bassanio, these people that lived 400 years ago, in a Facebook profile. I made them do

likes, dislikes, and they had to use passages from the play to show those things.

In creating Facebook profiles for the characters, the students were able to identify with them as real people, thereby helping to make the text more accessible and understandable. Martha, another EOP English instructor, used Adobe Reader to facilitate an in-class exercise in peer review of drafts of their papers. Instead of providing hard copies of the drafts, the instructor projected a PDF draft on the screen and then had one or two students annotate it in real time while the rest of the class provided collaborative feedback. In reflecting on the benefits of this method, she stated:

> I actually had the students do the annotations on Adobe Reader . . . and I think in some ways it made them feel more responsible for doing a really thorough job, marking down everything that anybody said. It was helpful in terms of pedagogy just for students to see this is someone's paper that I think is really good in some ways, but I can recognize that maybe they're all over the place, and we can help them figure out how to organize their thoughts and restructure it in some ways. They developed, I think, a more critical eye about their own writing when they started to see how they could help other people.

As in the case of the Facebook character assignment, the instructor's intentional use of technology in this example provided a level of engagement that would have been unachievable without the incorporation of that specific technology. Furthermore, the faculty felt that they had effectively achieved the intended student learning outcomes and engendered more student enthusiasm for the subject matter.

In contrast to educational technologies that simply render course materials more physically accessible, such as PowerPoint presentations, lecture recordings, or e-books, these strategies facilitate greater student interaction. Such implementation of technology represents the upper end of the technology use spectrum, but as our students indicated, even limited incorporation of technology in the classroom provides a more engaging experience. At minimum, faculty can support their students by allowing them to exercise self-determined strategies for enhancing their learning, such as taking notes with Evernote or recording the lecture, rather than forbidding the use of technology in class.

FACULTY AND STUDENT CONNECTIONS ON SOCIAL MEDIA

> So depending on what it is I feel like there might be that connotation of, if we use the Facebook page, it's a little bit less formal and official. I don't know if that's appropriate. I don't know if either party will feel it's appropriate for that function, for those particular functions of advising, homework, putting syllabi up on the page.
>
> —Mark

In addition to accessing tech-friendly course materials, such as PDF readings or e-books, instead of paper-based materials, students reported that outside class the technologies that their instructors used to interact with them were primarily course management software (e.g., Blackboard or Canvas) and email. Whereas course management software was typically limited to faculty-to-student communication with faculty sharing course information or uploading supplemental materials, both parties initiated contact through email. In general, students preferred face-to-face (F2F) dialogue with their professors, but when this was not possible or for simple questions, students viewed email as the most appropriate and professional means of communication. Krista explained her thought process for choosing how to interact with her professors: "Sometimes I like meeting with the professor if I'm discussing my ideas for a paper. But then I would just prefer emailing them if I have a question. Well, actually I just email them to set up meetings, or if I want to clarify, like a quick clarification question on the homework. But if I need help on the homework, then I would just rather them help me with me in the room." This decision-making process reflects most students' preferences for using email to ask faculty quick clarifying questions or to schedule a time to meet outside regular office hours. Email and in-person contact were also the most frequently used methods for communicating with teaching assistants (TAs). However, since TAs are often closer in age to the students and viewed as more approachable, students sometimes felt comfortable calling or texting them, especially when the TAs established that such forms of communication were acceptable.

Beyond social media apps, the most formal means of communication on campus occurred through emails with campus administrators and academic departments. Although emails were often referenced as part of the student experience of being at Millbrook, FGCS described email as a tool that was appropriate and useful only for official business purposes. This characterization was verified by Krista: "I know I don't really check my email that much. On the weekends I don't at all. Email for me is just for talking to professors or doing, I don't want to say important stuff, but professional stuff, I guess."

Unlike social media, where a casual approach to communication and interaction occurred, students stated that email was the less preferred means of communication on campus and its use was reserved to coordinate or schedule meetings with faculty or send questions to the TA of a class. This perception of email as strictly a formal means of communication reflects the importance and identity of social media as the mainstream virtual lifeline of student culture on campus. While campus administrators and departmental offices may have inclinations to engage with students on social media, these findings suggest email is students' preferred means for communicating with campus professionals and faculty. Social media, in students' minds, are reserved for connecting with students on campus, and not with faculty and administrative staff.

Our FGCS and faculty alike viewed contact through social media as unprofessional, emphasizing such technologies as personal and private rather than academic. Several students expressed the idea of social media as an escape from school and, therefore, opposed using them for academic purposes in order to prevent a disruption of this function. Meena, for example, stated, "If academics were to take over Facebook, or if they try to make it something more revolving around academics, I feel like, I don't know, I feel like I would lose that joy or lose that escape." Furthermore, both parties expressed concern about possible invasions of privacy in mixing social media with academic purposes, especially with regard to communication through Facebook. Mark, an EOP administrator, stated, "I definitely see Facebook as more of like a friend/peer media as opposed to kind of interstatus media." Students agreed that connections through social media could complicate the teacher-student dynamic. Camila explained, "I just feel like you're my teacher, you're not my friend. If I was a teacher, I would feel like I'm your teacher, I'm not your friend, let's not get that relationship confused." Despite the fact that both parties advocated the necessity of maintaining the boundary between personal and professional, they conceded the possibility that this gap might be bridged after the course ended if a closer relationship had developed.

Although most students acknowledged this boundary and professed the importance of being professional in their interactions with faculty, a few expressed an openness to alternative modes of communication if their professors designated them as acceptable. Nico expressed his thoughts about connecting with faculty through Facebook: "I don't want to have to add them unless they from the start they were like, oh, yes, add us. That's okay. That's how we talk. I prefer Facebook. If they said that, then I'd be like, okay I'll Facebook you. Most teachers keep that email boundary." Several FGCS did mention faculty who offered virtual office hours or student meetings through Skype or other video platforms, and their reaction to these options was mixed. While Yvette refused to participate in her biology professor's Skype office hours, viewing the prospect of connecting with her teacher at home as weird, Gabriel appreciated the immediate conversational advantage of the question-and-answer session that his writing seminar professor arranged before the final exam. The few students who advocated for a greater faculty and staff presence on and usage of social media highlighted the potential expediency it would allow for communication and the greater likelihood that students would view the information as compared to email. Still, it was clear that such a shift would have to be faculty- and staff-driven in order for most students to feel comfortable using personal social media for academic purposes.

In the few instances where faculty created Facebook pages for their classes, the platform provided an informal space for course-related group communication. Students could ask questions of the instructor or each other regarding homework assignments and easily discuss group projects. Such pages also allowed everyone to maintain the privacy barrier since group membership did not require sharing of personal profile information. Martha, who implemented a Facebook page for her class, highlighted the importance of the structure she instituted in getting students to use the page. "I think it definitely, when other students see that other students posted on it, it seemed more like it is okay to do this. It's okay to ask a question like, how do I cite whatever? I think as much as whatever I try to set up, that winds up having a pretty powerful effect." By building more of the course information on Facebook, even having students sign up for office hour appointments through the group page, she established Facebook as a normative method for communication. As a result, students utilized it more effectively as a course resource. Although the page traffic ceased after the class ended, the faculty member was still able to use it to informally reach out to her students once the fall semester began to check in and offer continued support.

Regardless of the setting, the most common behaviors associated with faculty interaction included discussing grades or assignments and asking faculty for information related to a course with 45.4% and 39.6% of students, respectively, reporting engaging in these behaviors a few times a month to at least once every week. Speaking with faculty about career plans occurred with the least frequency, with 18.9% of students never doing so and an additional 28.3% discussing their careers with faculty only once during the academic year (Table 5.2).

Differences also emerged with regard to how comfortable our FGCS felt in communicating with instructors. The average

TABLE 5.2. Student engagement in certain activities to connect with faculty (n = 53)

	NEVER	ONCE DURING ACADEMIC YEAR	A FEW TIMES A SEMESTER	A FEW TIMES A MONTH	AT LEAST ONCE EACH WEEK
Visited informally with faculty before/after class	13.2	15.1	43.4	13.12	15.1
Made an appointment to meet faculty in office	3.8	11.3	47.2	32.1	5.7
Asked faculty for information related to a course	3.8	11.3	45.3	22.6	17.0
Discussed grades or assignments with faculty	7.5	15.1	32.1	34.0	11.4
Talked about career plans with faculty	18.9	28.3	30.2	9.4	13.2

student reported that the ease with which students engaged in all activities connected to faculty declined over the course of the year. With the exception of making an appointment to meet with an instructor, the students' initial pre-entry rankings indicated significantly higher expectations for their anticipated interactions with faculty than their actual behaviors in August and March of the academic year. Thus, although students recognized the importance of developing relationships with their professors and desired this contact, their connections with faculty overwhelmingly remained weak ties. Their ties to some of their EOP faculty did take on the characteristics of strong bonds, however, and some students did obtain the psycho-emotional support from these instructors that is common to bonding capital.

When it comes to using technology for interaction with faculty, whether for academic purposes in the classroom or as a means of communication, students follow their professors' lead in establishing acceptable behaviors. Many of the EOP instructors referenced teaching the students how to appropriately interact with faculty as an important objective of the program, and they also recognized the importance of role modeling the use of technology for academic purposes. Lilly commented that "the students mimic the people they're closest with. So, I think seeing the academic people in their lives, the faculty members in math and English using it, in those academic ways, will make a bigger jump to how technology can help you, and help bridge." Not incorporating technology into the classroom, therefore, results in a missed opportunity to help students make stronger connections with course material by engaging them with the sophisticated devices that are part of their everyday lives.

Staff Interactions and Bridging Capital

In contrast to their more limited interactions with their instructors during the academic year, our FGCS continued to regularly connect with professional staff and with student leaders affiliated with the EOP. The students reported strong relationships with a variety of individuals, including advisors, counselors, peer mentors, Office of Multicultural Affairs (OMA) staff, library staff, and the staff instructor of a college transition course. Several students had nothing but positive associations with affiliated EOP staff, even months after the program concluded. Mason "really [looks] up to them actually," while Nico described a consistently positive relationship with his counselor in which "it's like nothing has really changed" since the summer and during meetings they "always have a good time." Meena described an especially close relationship with her former Peer Mentor: "I call her my sister now."

For most of the students in our study, the connections with these staff are strong and consistent and have the characteristics of both bridging and bonding capital. While a plurality of the sample (47%) reported interacting with their advisor a few times each semester as the program required, nearly 10% reported interacting with their advisors at least once each week, and another 11% reported similar connections a few times per month. Along with offering general problem solving and connections to academic resources, professional staff and student staff help students gather important information that can be utilized as bridging capital, in areas that include course selection, financial aid, major declaration, campus events, and room selection (Table 5.3).

Nico told a story about how the staff instructor of a college transition course encouraged him to visit his professor's office hours: "Last semester I think I met with Mark about the problem, and he was like, oh you should meet with the professors more and get to know them and they can help you a lot. I was like, maybe I should. Then spring semester I went to I think a few office hours from each class, and I ended up liking it. So I just started to go more."

Perhaps unsurprisingly, these strong relationships are forged and maintained in an offline environment well suited to the professional staff who establish the bridge program and advising structure. An advising requirement—students must visit their advisor three times each semester—sets a clear tone; in order to connect with a supportive mentor, students must relate F2F. However, students who desire a closer connection must elect to stop by their advisor's office or some other physical location (e.g., the OMA) where professional staff work. For some students this system works effectively. As Gabriel described, "I just pop into the office, and they're very open to help me about new things, scholarships, the

TABLE 5.3. Percentage of respondents who reported using various formats to learn about processes on campus (n = varies)

	EMAIL	WEB	FACEBOOK	CLASS	OMA STAFF	MU STAFF	RA	IN PERSON
Registering for classes	76.1	30.0	25.8	57.9	48.7	75.0	56.8	83.0
Financial aid	84.4	48.6	10.7	10.3	52.8	66.7	25.0	47.4
Declaring a major	54.3	29.0	3.6	14.3	51.4	77.8	24.4	50.0
Millbrook events	78.7	47.1	76.1	53.8	35.5	41.2	55.6	76.1
Housing selection	69.4	37.8	38.9	23.5	14.3	30.3	66.7	71.7

tuition admission application, and all that stuff, and they email me, so I think it's really good." Another student described feeling "more comfortable going into the office of the OMA" than attending an event. Still, other students cited a busy daily schedule—one often incongruent with staff offices that operate on business hours—as the reason why they did not interact with well-liked advisors more. Some accepted this on-again, off-again relationship and emphasized the positives like Yvette: "They've all been really supportive. I just haven't been able to keep contact as much as I would like to because I'm so busy all the time. But every time I see them, they just give me a big hug, catch up on everything real quick, and it's good." Others would prefer more outreach. Krista shared that "unless you stop by [the] OMA you'll never hear from them unless they're sending you their newsletter." To be sure, this is not always for lack of trying. In conversation about EOP Facebook pages, a group of peer mentors acknowledged that students generally do not spend much time on social media pages created by non-students with Colin sharing, "when students create it, it's different. I don't know why."

The relationship between staff and students changes slightly when examining connections with peer mentors and advisors. As students at the same college, the student leaders were able to ally with faculty and professional staff during the EOP and then shift to a peer stance once the academic year began. And while in-person interactions remained consistent during and after the EOP, technology appeared to provide the crucial pivot point upon which peer mentors could enact different roles in relation to their advisees. Specifically, while acting as peer mentors they often refused to connect with EOP students over social media. One peer mentor, Colin, a self-described "big Twitter user," was "actually following a student, and that person followed me back. We had to like block each other, just for the program, but I told him, once it's over, I'm going to unblock you." This pattern held across multiple interviews: once the program ended, the walls came down; peer mentors and students could follow each other on Twitter, friend each other on Facebook, and exchange Snapchats. Ruby defended this relational about-face bluntly: "I don't see why not, because then we're both students and we're both at the same level, we're both adults."

Bridging Capital Attained Through Weak Ties

Our FGCS attained bridging capital through weak ties—relationships with faculty, staff, and peers that are less intimate and more detached than those found with close family and friends. FGCS in this study were often eager to meet with their professors and mentors F2F outside of class. Office hours and email, however, seemed preferred communication means for FGCS and the purposes of these communications were primarily transactional, that is, to gather new information. Their F2F behaviors described were savvy and goal-oriented: sitting in the front row, saying hi to a professor after class, and stopping by during office hours to discuss homework or an upcoming assignment. One student noted, however, that creating personal connections with faculty depends on faculty's approachability: "the way that the professor portrays their personality or their availability, sometimes makes you either come up to them or either deter yourself from going up to them." Initiating F2F communication with faculty, according to these FGCS, hinged on students' perception of faculty's approachability and receptiveness.

In examining the breadth of student experiences across the modern university, it becomes clear that the nature of the academic enterprise lends itself to knowledge transmitted almost exclusively across weak ties. For instance, students who make the effort to connect to faculty members may do so only for a single semester for course-related purposes. Some EOP faculty, many of whom forged strong relationships with their students, may no longer be on campus once the academic year begins. Across the institution, many faculty and staff avoid engaging with technology beyond email, further limiting their availability and opportunity to connect with students who spend much of their time using social media and mobile devices.

The students who spoke of deeper relationships with faculty or other mentors managed to break away from the typical communication conventions. For instance, some students chose to meet with faculty with whom there was no current relationship (i.e., not taking a class presently with this person), and with no particular purpose in mind beyond conversation. Others described meeting up with faculty for coffee or dinner, using technologies like Skype and texting to facilitate a stronger connection with their EOP peer mentors, and building rapport through regular interactions within a living-learning community. When the typical evaluative relationship between faculty and students diminishes or comes to an end, students can develop stronger relationships with faculty. Camila explained when describing her relationship with Lilly, an EOP faculty member: "I mean there is still like that respect and authority figure obviously above me, but I feel like it's easier for me to interact and talk to her when there is more of like a personal connection as opposed to like an instructor/student connection." For some students, these deeper relationships were conceptualized through the language of mentorship. Academic advisors, EOP faculty and peer mentors, and staff at the OMA office were experienced by students as mentors because they were available and gave reliable advice. Carolina captured the closeness she felt to Martha, a faculty member who had been her instructor during the EOP and through the subsequent two semesters: "She is like my mentor here at Millbrook, so I really enjoy having her around. So I like that [Martha] is always there if I ever need help. She is not my academic advisor, but if I need help she is there, and I know I can go to her if I ever have an issue with anything, so I like that about her, and plus she has a daughter who is like my year, so she understands what is happening."

This chapter captured the ways that students regularly rely on social media in order to acquire the necessary knowledge that will aid them in their social success and persistence on campus. College administrators and those invested in working with FGCS would benefit by increasing their awareness of how FGCS interact on social media, considering how technological tools can be leveraged for the promotion of student success, and contextualizing how social capital can further be circulated within the virtual spaces that FGCS regularly frequent. Attention to these means would better prepare and equip FGCS with the necessary tools that advance social integration, promote campus capital online, and further encourage students to be immersed within a campus environment that is designed to facilitate positive experiences at all levels within higher education. We also examined the at times rocky relationship between students, faculty, and the use of technology in the classroom. When faculty used technology in the classroom, and promoted student use, there was a benefit in terms of what students were learning related to content and the technology skills that students learned along the way. These practices helped students to be more engaged in the material, and ultimately to feel a greater sense of belonging in the classroom. Unfortunately, innovative uses of technology were the exception, not the rule. Faculty did not often have the skill set or capacity to integrate new forms of technology in their classes without additional support. We examine some of our conclusions and recommendations in the following and final chapter.

6
Propositions for Change

WITH CONTRIBUTIONS BY KEVIN GIN

As Professor Toby, an award-winning faculty member recognized for his use of technology in the classroom, finished describing the final group assignment, Yvette could feel herself growing excited. The class was being asked to research the history of a specific location and create a game where a visitor could learn about the place. Yvette thought to herself, "THIS I can do." First, she loved history. Also, she used to play games all the time with her cousins at home and could imagine playing a game about history. Plus, the opportunity to work with Arun and Camila was a huge bonus. She knew them both from the Educational Opportunity Program (EOP) but did not have a chance to get to know them well. She was certain, however, that working on this project together would help her connect with them. Plus, the fourth student in the group, Brian, seemed like a great guy. He was smart and always seemed to know everything about Millbrook. On a few occasions, Yvette had asked him for suggestions and he was always more than generous with his time. Although Yvette had never spent time with Brian outside class, and did not think Arun or Camila had, she was psyched that he was in her group too.

Professor Toby told the students that they could use the last 15 minutes of class to gather in their groups and make a plan for completing the project. Immediately, the four students began swapping cell numbers so they could set up a GroupMe and Gmail addresses so they could Google Chat some of their meetings. Then, they began to brainstorm possible places. Together, they decided they would focus their project on the history of Millbrook. It was brilliant! They could learn about the history of a place in which they were actually living. And creating a game about it was something they could give to new students. Brian added that he thought alumni might even play it.

As Professor Toby asked students to wrap up their conversations, Brian suggested that they start their project by meeting together with a reference librarian. "What a great idea," Yvette thought. She knew she would not have suggested it because although she had heard of those staff, she was not entirely sure what they did. As the group packed up their iPads, they agreed to individually read the assignment description Professor Toby put on the course Learning Management System (LMS) and then text possible meeting dates. Yvette felt herself skipping back to her room thinking, "Okay, I can do this."

For Yvette and the other first-generation college students (FGCS) at Millbrook University, going to college was a huge accomplishment for them and their families. As the first person to go to college, attending Millbrook was a key step toward a fulfilling future. However, this milestone brought academic, financial, social, cultural, and emotional challenges. Knowing these challenges are persistent for FGCS at college campuses around the country, we began this project with the belief that connecting FGCS to faculty and peers through Web 2.0 technologies (social media and iPads) would address these challenges and in doing so enhance the relationships and experiences that are critical for student success. Our goal was to provide a scholarly view of how college programs dedicated to improving first-generation college students' experiences could utilize technology effectively. Based on what we learned from students, staff, and faculty at Millbrook, we contend that social media and Web 2.0 technology can serve as an equalizer for FGCS to engage on campus by providing an easily available means to acquire college capital necessary for successful social and academic transitions. As we saw, for students like Yvette, Nico, Carolina, Arun, and Camila, technology acted as a bridge to essential information, relationships, and opportunities that will be critical to their ability to graduate from Millbrook.

In this chapter, we summarize the major conclusions we identified through our project and share recommendations for practice and research that were informed by our project. Also, we revisit the conceptual model presented in the Introduction and propose a revised conceptual model that can be used to develop technology-based practices that will positively affect FGCS success on campus and research that seeks to further our knowledge of FGCS college engagement and success.

Conclusions

PROMOTING A SUCCESSFUL TRANSITION

The emphasis on technology and social media's presence in the experiences of FGCS has become a central component in the promotion of successful transitions in college life. The accessibility of technology not only helps increase access to campus capital, but these virtual and mobile tools further facilitate the transition to college by promoting a sense of community, ensuring engagement in campus life, and strengthening relationships with home (strong) ties.

Students repeatedly noted and gave priority to their connections with family and friends from home through smartphone calls, video chats, text messages, and social media. These strong ties are vital sources of psycho-emotional support that have been previously noted to be crucial to the persistence of FGCS in higher education by positively affirming the ability of students to succeed in culturally unfamiliar environments (Granovetter, 1983). The findings from our study, highlighted in Chapter 4, are evidence that social media permit the continual maintenance of these central relationships across considerable distances and disparate university/college settings through convenient, low-friction technologies such as text message/FaceTiming and social media platforms such as Facebook and Snapchat. These virtual connections enabled students to regularly access a window into the lives of their peers and family through texts, photos, and videos, and connect in meaningful ways to remediate the stress that often accompanies the transition of FGCS into college (Orbe, 2004).

While strong ties comprise the unique and irreplaceable inner circle of FGCS networks that mobile technologies and social media helped maintain throughout the college experience, students were also reliant on their weak ties at Millbrook for social connections and the acquisition of campus capital. Weak ties contributed to the integration of FGCS into the campus setting in important ways, including through virtual communications on social media platforms that are central to today's student experience.

Indisputably, social media such as Facebook, Instagram, and Twitter, and newer apps like GroupMe and Snapchat play a key role in shaping the culture of today's students by deepening connections and circulating campus capital between FGCS and their peers. Facebook continues to serve as the virtual kiosk that enables these outcomes through the promotion of extracurricular opportunities, campus programs, and connections with weak ties. As noted in Chapter 4, Facebook enabled students to engage with weak ties and the campus setting in a number of ways, including disseminating information regarding upcoming involvement fairs, publicizing cultural shows where peers were performing, and promoting new friendship groups through connections with student organizations. The virtual means that campus culture typically co-exists with face-to-face (F2F) interactions on today's college campuses and speaks to the hybrid nature with which higher education must address the relationships between students and their peers in the twenty-first century (Martínez-Alemán & Wartman, 2009). As a result of being constantly connected to their peers and campus life through mobile social media, today's FGCS appear to have a stronger hold on connections to both strong ties and weak ties in ways that prior generations did not have due to the absence of these virtual connections. The reality of today's students co-existing on both physical and virtual campus cultures raises questions and concerns regarding the corresponding barriers that threaten the positive outcomes enabled by social media on today's campuses and necessitates further scholarly and administrative attention.

Previous researchers noted, and this study confirmed, that racial tensions on a predominantly White campus tend to promote feelings of isolation and loneliness among FGCS (Sarcedo et al., 2015), but social media effectively and instantaneously connected students with culturally familiar campus peers within ethnic communities at Millbrook. In response to the presence of racial hostility both F2F and online during this study, students were able to leverage social media as tools to mediate these tensions, and engage in wide-reaching communications with likeminded peers through virtual connections and expressions of defiance against racial inequities (e.g., Krista championing #BlackLivesMatter as an ongoing dialogue on social media and on campus). The convenience of instantaneously connecting with an online community of peers as a site of resistance during experiences of racial duress distinguishes FGCS on today's college campuses from previous generations of students who were without these online tools and faced less convenience organizing their efforts to mobilize communities in response to campus racial hostility (Altbach & Cohen, 1990).

Today's FGCS are faced with remediating the cultural divides that emerge in both F2F interactions and virtual environments, but the versatile functionality of what social media can accomplish on today's college campuses (e.g., constant connection to others, virtually identifying empathetic and compatible communities, capturing and sharing injustice instantaneously with others) exemplifies the potential that technological advances can promote for students seeking culturally familiar and safe environments on campus. For example, social media permit students who encounter localized issues of inequity on campus to amplify those experiences through hashtag activism (Bonilla & Rosa, 2015); or to identify and connect with allies in curated online spaces where privileged conversations may take place with trusted peers safe from a hostile F2F setting (Florini, 2014; Sharma, 2013); or to further establish and strengthen connections with friendship networks, including both weak and strong ties (DeAndrea et al., 2011). In the instances of our students, social media enabled FGCS to exhibit a combination of all these outcomes, and simultaneously foster an increased sense of identity to ethnic subcommunities at Millbrook, which were effective in resolving feelings of loneliness and promoting a stronger sense of belonging. These outcomes, which were specific to the students in our study, exemplify the constructive possibilities that social media and technology are capable of facilitating for FGCS.

Overall, it is increasingly important for administrators to develop a presence in the social media spaces that students frequent, both to understand the racial hostilities that many FGCS experience online and to advance competencies for intentionally engaging FGC students in social media that facilitate the dissemination and acquisition of campus capital. Currently, administrators minimize the role they can play on social media and often do not look to social media to dispense campus capital to FGCS. This is despite testimonials from participants in our study that recommended Millbrook administrators more intentionally use social media to advance constructive student experiences stemming from a highly foreign environment (e.g., Yvette advocating for administrators to use Facebook instead of email), and calls by researchers for higher education leaders to leverage social media to facilitate institutional social change (Gin et al., 2016). Many administrators mistakenly perceive social media as solely the private, social domain of students or as a space for other institutional functions. Instead, administrators should understand social media as a space in which they can execute co-curricular education.

ACADEMIC INTEGRATION

Our FGCS gained academic knowledge and bridging capital predominantly through weak ties with faculty, staff, and peers. Web 2.0 technology can facilitate various aspects of this knowledge acquisition process. Specifically, the intentionality with which faculty, staff, and other mentors introduced, supported, and encouraged students to use technology appears to have impacted its effectiveness for learning. Premeditated and purposeful use of social media by faculty, staff, and mentors—whether to distribute campus capital as academic knowledge or institutional information—mattered greatly to students. Students responded to well-designed use of social media and tablet technology in the classroom; they perceived social media as useful learning tools that engaged them in class activities. In academic environments where technology was used effectively to enhance learning and engagement, FGCS felt a greater sense of belonging in the classroom. The findings from our study further resonate with and confirm previous research that has shown that intentional and targeted implementation and usage of social media within academic settings results in positive outcomes for student learning and engagement (Delello, McWhorter, & Camp, 2015; Hamid et al., 2015).

Our FGCS expressed an eagerness to use new technologies in the classroom and voiced their frustration with courses where it was not allowed. This dynamic at Millbrook exemplifies the evolving tension that exists regarding the question of how higher education can most effectively and intentionally reconcile the silos between the virtual and F2F cultures that exist on campuses today. Although our study confirms that FGCS positively receive the introduction of online forums and social media in the classroom, this reality is confronted by difficulties regarding how to integrate this technology across diverse academic programs and faculty with varying degrees of technological competency. The most common technology used in Millbrook classrooms by faculty are PowerPoint slides, video presentations, and course management software, which are the most simplistic and minimal technological tools. Students view faculty who fail to use these platforms as anachronistic, especially if outmoded technology, such as chalkboards or overhead projections, are employed instead. Mobile apps, social media, and video conferencing appear less frequently, but when introduced by faculty in the context of a course, students reported positive experiences. In general, classmates obtained information about academic processes from one another in person but less frequently through social media. However, social media and mobile apps were regularly used by FGCS to support collaboration with classmates and class-related group work or to make requests for specific course information, such as notes or assignment clarification.

These findings affirm that the purposeful incorporation of technology and the encouragement by faculty to use social media within classroom activities (e.g., lectures, group projects, individual assignments) fosters an academic environment that is conducive to the acquisition of knowledge, and solidifies the communal means by which learning can occur on today's hybrid F2F and virtual college campuses. Similar to the ways that institutional leaders have resisted integrating social media within administrative processes to advance campus social life (noted in Chapter 4), faculty who are averse to embracing technology in their processes and/or do not possess the literacy to navigate social media for academic purposes do so at the risk of inhibiting the promotion and maximization of learning in the twenty-first century.

Within the scope of this study, FGCS profited from the presence and integration of technology and social media in ways that permitted them access to increased learning resources through online searches; convenient compilation and review of classroom material in transportable tablet form; and the ability to instantaneously capture knowledge through digital imaging, audio recording, and annotating academic content in ways that liberated students from single-dimensional pen-and-paper note-taking. As exemplified by the academic benefits that our students received from these previous examples of online resources in the classroom, the ubiquity of technology and social media on today's college campuses (Duggan et al., 2015) necessitates the endorsement of virtual academic tools and mobile accessible apps that effectively address the needs of increasingly more technologically fluent FGCS on today's campuses.

Despite the plethora of benefits that are possible through the previously described practices, students were clear that the use of social media by faculty could not transcend the academic space. In other words, students want faculty to use social media in classes as long as their privacy is not breached. This response by students embodies the complications that faculty and academic administrators continue to face in their assessment of how to best promote, use, and seamlessly incorporate social media as educators. Our students recognized the pedagogical value of social media by creating weak ties with faculty within the context of course assignments. However, students were clear that they wanted to preserve social media as *their* social spaces, as spaces for their co-curricular *social* lives.

Students' weak ties with faculty and staff were largely established F2F or through email communication but not through social media. Students did not connect with faculty and staff through Web 2.0 technologies, instead opting to employ email, a twentieth-century technology. Most students seemed to accept that email communication with faculty and staff was standard on campus, albeit it represents an older and restrictive technology. They adapted to this preexisting structure and custom, even when it did not fit the contours of their everyday lives as effectively as Web 2.0 and mobile modes of communication might. Students did not employ Skype or FaceTime to communicate with faculty and staff, nor did faculty and staff regularly offer online office hours. Among our FGCS there exists a belief that email communication with faculty and staff is a convention that is appropriate given that these are not "personal" relationships with these adults in positions of authority. Our FGCS maintained these weak-tie relationships as impersonal and did not believe that social media were a means to communicate with faculty and staff. Ironically, FGCS accepted faculty social media use as part of class activities—as learning tools—but they did not employ social media as a relational or social technology with faculty.

Nonetheless, both FGCS and faculty were open to broadening their use of Web 2.0 technology across the academic sphere and would welcome additional training in order to do so more effectively. A small number of our FGC students did report deeper relationships across academic weak-tie networks, utilizing Web 2.0 technologies (e.g., texting, Skype) alongside more conventional offline vehicles (e.g., in-person meetings, dinner at a faculty member's house). One possible implication of this finding is that faculty and students are certainly able to facilitate a greater accumulation of bridging capital through social media but since so few did so, perhaps faculty require greater awareness and more professional training across a suite of technologies to enhance academic relationships for a greater number of students. Faculty may well need to take the lead in normalizing communication with students on social media without losing sight of propriety and the ethical boundaries of their respective roles.

These nuances regarding the boundaries of how social media can be used to best support learning and faculty connections further emphasize the need for higher education to explore in more depth the practices and expectations that best facilitate academic success (including how to best connect and maintain relationships with faculty outside the classroom). In the instance of our study, FGCS see the possibilities of using technology and social media as a means that further enhance, promote, and offer opportunities to engage in academic achievement, but also where distinct boundaries exist between the junction of their social and academic spheres within the student experience. Although our findings demonstrate that the integration of technology and social media within academic experiences results in beneficial outcomes to the learning process, incorporating these tools in practice requires a conscious purposing of social media and technology by faculty and staff that is narrowly tailored to align with the distinct expectations and needs of today's FGCS (i.e., social media must be purposed in a way that straddles the line of being a convenient and ever-present tool used by faculty to supplement learning in the *academic domain* without crossing into the territory of encroaching on a student's online *personal domain*).

THE POWER OF PEERS

Peer mentors occupied a unique space in this study. In their official capacity, peer mentors behaved more like faculty or staff and maintained stricter boundaries around interaction with our FGCS across social media platforms and texting. Once their formal role ended, however, the peer mentors used these Web 2.0 technologies to facilitate their transition to fellow undergraduate students. This modification of their status—peer mentors moved from being quasi authorities to campus peers—enabled mentors to establish stronger weak ties with FGCS. Once peer mentors became fellow undergraduate students, FGCS sought them out on social media and strengthened already established weak ties with them on social media. The fact that these former peer mentors shared many of the same characteristics as FGCS—many were FGCS and students of color (SOC) themselves—permitted their ties to develop characteristics of strong ties, yet maintain their weak-tie character (Daniel Tatum, 1997; Fleming, 1985; Gilbert & Karahalios, 2009).

The relational ties through social media between FGCS and the former peer mentors deepened their weak ties but did not prevent the circulation of bridging capital. That is to say, former peer mentors could still transmit campus capital as a function of their weak-tie status but because they shared so many social commonalities with FGCS, they could also circulate bridging capital—they could communicate as strong ties. These ties have a Janus quality. They are relationships that guard beginnings, transitions, and progression for FGCS by disseminating campus capital as both weak and strong ties would. Former peer mentors can communicate campus capital to FGCS about the politics of service trips on campus, graduate school test preparation, dating in a Predominantly White Institution (PWI), or paid summer jobs on campus. Concurrently, former peer mentors are sources of emotional and social support and validation. Like campus capital circulation through strong ties, campus capital disseminated by former peer mentors is intended to shore up FGCS insecurities, self-doubt, and anxiety about belonging on campus.

While the interactions between peer mentors and FGCS did not exclusively exist online, the nature of these evolving relationships demonstrates the importance of social media as a virtual forum that has tangible benefits. As presented here, relationships between FGCS with weak and strong ties are malleable and can evolve over the course of college experiences. The intensity of these relationships (Granovetter, 1973) evolves through a number of interactions that are augmented due in part through social media communications. These communications by students in our study, through both F2F and online, have the potential to reinforce relationships with weak ties in such a way that these initially heterogeneous connections may develop into culturally familiar strong ties that are integral to reinforcing the belonging and self-efficacy that are necessary to succeed on college campuses. It is because of this potential for FGCS to develop deeply trusted inner networks from their weak ties that increased attention must be dedicated to how technology and social media can possibly support the development of information-rich peer networks, and the augmentation of strong ties that are critical in fortifying the individual qualities that cultivate positive experiences associated with student success.

Social media can effectively accomplish the previously outlined objectives in this conclusion by circulating capital between FGCS and their peers on college campuses. The acquisition of campus capital is critical for promoting student engagement and success in higher education, especially for FGCS who are less likely than their continuing generation peers to be transmitted knowledge from their parents about the social behaviors, campus resources, and cultural norms that contribute to successful college experiences (Armstrong & Hamilton, 2013; Cox, 2011). The presence and intentional usage of social media as a means to disseminate this capital further makes accessible the campus knowledge that has been shown to be effective in facilitating equity for FGCS. Because of the numerous benefits regarding campus capital that emerge from social media, we propose a model that can be used as a framework to explain and comprehend how such capital is both transmitted and received by today's FGCS.

WEB 2.0 TECHNOLOGY AND FGCS: A NEW MODEL FOR TRANSMITTING CAMPUS CAPITAL

As an intervention intended to provide FGCS with another means to access the social capital necessary to succeed in college, our project sought to make use of the benefits of Web 2.0 technology and social media. We reasoned that because social capital on campus, what we termed "campus capital," is circulated through relationships the nature of social media and tablet technology was consistent with this goal. Understanding that social media could provide another means for FGCS to amass forms of campus capital, especially the bridging capital in weak ties, we set out to examine how FGCS would use social media and tablet technology in this way. We had anticipated that FGCS would extend their relational reach on campus to more and more weak ties, thus accessing more and more campus capital, and that their strong ties with home associations would remain salient. In ways, both of these presumptions were imprecise.

Though we certainly assumed that strong ties with home communities would be salient for FGCS, we had not anticipated their unique significance and how that significance was heightened by social media and tablet technology. Scholars have examined family achievement guilt (Covarrubias & Fryberg, 2015), perceived and maintaining family support (Kennedy & Winkle-Wagner, 2014), and community cultural wealth (Yosso, 2005) among FGCS and SOC and, not surprisingly, have confirmed the value of home communities for these students. Where perhaps among White, more affluent continuing generation college students (CGCS) helicopter parenting may be detrimental to those students' college transitions and developing autonomy (Schiffrin et al., 2014), among FGCS strong ties with parents appear to help them sustain their commitment to college (Nichols & Islas, 2015). As Nichols and Islas (2015) have shown, CGCS parents pull their children through their first year by giving them the information/social capital necessary for success. FGCS are pushed by parental support or bonding capital that through Web 2.0 technology can now be more easily accessed and thus more efficacious. At least, it appears so in the case of our FGCS.

As shown in Figure 6.1, our intervention with Web 2.0 technology and social media enabled FGCS to reinforce strong ties and consequently to acquire the psychological and emotional capital necessary to persevere through trying times at college. Easier and frequent contact with home communities augmented the bonding capital that fortified FGCS in their efforts to endure challenges and uncertainties, and to bolster their confidence and self-worth.

However, the intervention did not expand FGCS relational reach to weak ties that historically contain bridging capital. FGCS

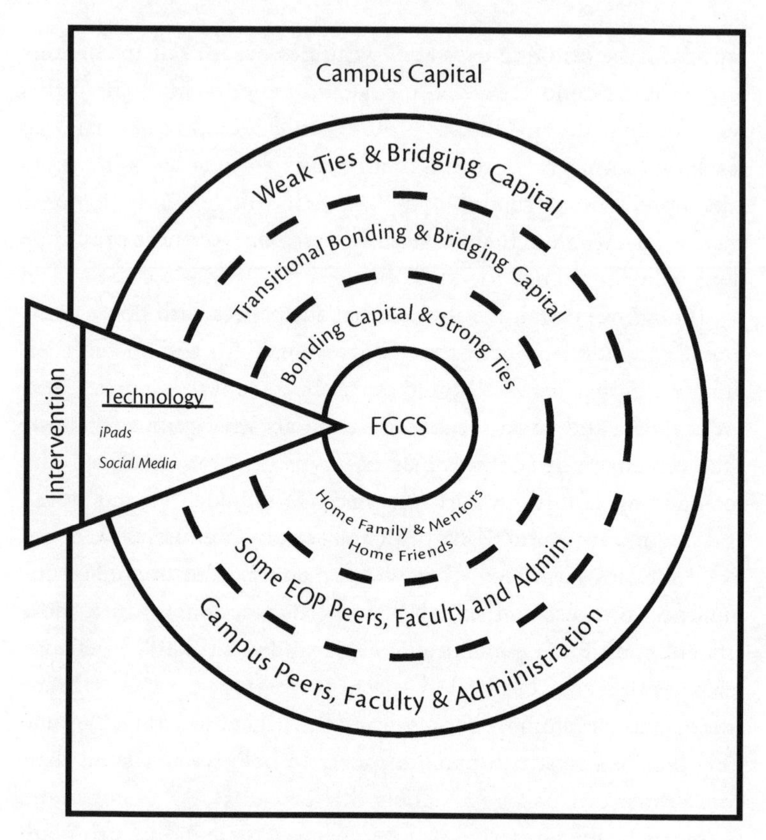

FIGURE 6.1. Model of Campus Capital for FGCS

extended their reach to weak ties by creating new ties with other FGCS or students of color, many who were enrolled in the EOP. These ties began as weak ties, that is, they were new relationships undefined by a previous history. However, because these relationships were with students much like themselves—they were FGCS and often SOC—these ties developed strong-tie characteristics and became sources of bonding capital. These ties with other EOP peers, former peer mentors, faculty, and staff were sources of a synthesis of bridging and bonding capital; this is capital we call "transitional capital." It is capital that FGCS sought out and obtained during their college years but that may not be maintained after graduation. Transitional capital appears to be an essential and critical type of campus capital that social media technology enabled for FGCS.

Recommendations

Based on the FGCS experiences described in our project, we believe there is room for new practices that utilize technology in ways that are responsive to FGCS unique needs and that foster communication across strong and weak ties. To begin, given the role that social media play in the lives of college students, it is critical that higher education administrators leverage the unique benefits of social media to enhance FGCS experiences while remaining mindful of their technological and personal preferences. Yet, FGCS unique needs may run counter to campus practices. Whereas our research suggests that administrators should be wary of using Facebook to actively engage students, there is good reason to believe that students are interested in passively engaging with college-related Facebook pages. For example, program and department administrators might consider migrating departmental websites (e.g., Student Affairs, Orientation) onto Facebook to ensure that key information is meeting students where they are, that is, in virtual spaces. Many of the students in our study described gaining a lot of important information through social media. On the other hand, it is clear to us that FGCS, protective of their emerging identities and lack of campus know-how, will resist engaging publicly on these social media sites. In our project, FGCS feared posting questions or comments that exposed their lack of campus capital. Moreover, FGCS are managing a challenging transition across social groups and social classes. As such, public and active engagement in social media is risky. FGCS worry about how they will be perceived by CGCS, White students, and students from more moneyed families. This desire to manage impressions trumps any perceived benefit of publicly engaging on campus-facilitated social media. Put simply, our findings suggest that students will reject technology uses that run counter to their perceived social positions on campus and their developmental dispositions.

A second major takeaway from our project was the desire for faculty to intentionally integrate technology into their teaching. We imagine that there are campuses around the country that have more thoroughly and successfully incorporated social media and Web 2.0 technologies in the academic space. On other campuses, faculty may not have resisted the adoption of these technologies and have more effectively integrated these technologies into their pedagogy. However, this was not the case at Millbrook and in the EOP, where faculty resistance and sometimes outright prohibition of technology in the classroom closed opportunities for FGCS learning. However, FGCS may uniquely benefit from faculty usage in ways that level the playing field academically and socially. For example, our FGCS described the ways in which apps helped fill knowledge gaps in their classes. Likewise, faculty willingness to meet over Skype or through other mediated technology signaled an openness to meeting FGCS where they are most comfortable.

Encouraging this level of faculty use will require significant changes at the institutional level. First and foremost, technology must be accessible for all students. In addition to making technology available to students (both hardware and software), institutions will need to expand faculty/staff training and development focused on the integration of Web 2.0 technologies into their teaching. Indeed, this change might mean that departments of academic affairs need to incentivize faculty training and classroom usage (Martínez Alemán, 2014; Moran, Seaman, & Tinti-Kane, 2012). If faculty and staff are not comfortable with technology, and are not providing training and time to really become competent with these tools, faculty will be less likely to use them at the risk of not being able to meet the learning outcomes of the course, and/or to risk appearing unprepared or unknowledgeable in front of the class.

Currently, FGCS who may be struggling to learn the code of higher education may be further disadvantaged by differing policies and norms among the faculty. Institutions should also implement a clear policy on academic technology and educate FGCS on what this policy implies and really entails.

As institutions embrace the benefits of technology and prepare their faculty for classroom integration, we offer two recommendations. First, FGCS described how various apps supported their learning and classroom engagement. However, not all students have the same level of technological literacy. Thus, we imagine FGCS would, like faculty, benefit from training on how technology can enhance their academic experiences. While some students will need further training, others will enter classrooms with advanced technological skills. FGCS who had experience with apps or who were frequent users of Web 2.0 technologies knew how to engage with academic assignments in ways that students unfamiliar with these technologies could not. Some of our FGCS were not predisposed to explore the many different uses of apps or to extend their use to other courses or social needs. The more tech savvy and experienced FGCS were able to exploit the benefits of the technology for their course assignments and learning. Faculty in our study who integrated technology in their classrooms described instances in which students' expertise with technology outpaced that of the faculty member. This dynamic—where the student is bringing funds of knowledge potentially beyond that of the faculty member—is powerful and potentially effective. By allowing students to be more expert-like, there is a shift of power that takes place, which puts students in a good learning position. We believe this could advantage FGCS greatly; however, it may require a shift in thinking among faculty, who are used to maintaining the expert role in their classrooms.

One of the major conclusions of this study is that FGCS benefit greatly from weak and strong ties. Regardless of tie strength, each transmits important and valuable sources of campus capital. Given the value of these relationships, we suggest that institutions consider the types of opportunities that facilitate each type of tie. Certainly, FGCS weak ties with faculty, staff, and students with whom they do not share an affinity or interaction provide them access to information and resources that promote their success in college and beyond. In our project, students described how weak ties provide important sources of campus capital, especially instrumental or resource-oriented capital. For example, students noted how through weak-tie with peers they would learn about internship opportunities in the community. In order for this communication to happen, higher education leaders should ensure there are opportunities for FGCS to network with people within and outside their existing social networks, especially when those social networks are homogeneously similar to FGCS. In their study of sociology students, Spalter-Roth et al. (2013) found that career-focused capstone courses were an effective vehicle for students to network for jobs using weak ties. Thus, university or department-wide courses such as capstone courses or first-year courses where students are introduced to many types of campus capital (e.g., campus resources, opportunities for learning, support services) are uniquely positioned to foster weak ties. Conversely, institutions should monitor social networking programs that are too homogeneous in nature. For example, campus-wide programs solely targeting first-generation students can miss the mark in terms of creating a context for campus capital transmission.

Whereas weak ties are important to FGCS, our study also highlighted the benefits of students' strong ties, especially as students are transitioning into a new environment or taking on a new challenge. Family members and peers from home settings provide FGCS with an essential type of capital. These relationships offer social and emotional capital necessary to withstand the stressors associated with being the first in one's family to attend college. Thus, we recommend that those working closest with FGCS should more intentionally provide means for them to connect with their friends and families at home. Instead of only sponsoring parental involvement workshops and seminars to address the helicopter parents of CGCS, institutions can make technologies available to FGCS and their families that would enable better access to the bonding capital in those relationships. Institutions must seek opportunities to strengthen FGCS access to these strong ties, especially in students' first year through Web 2.0 technology. In an era where concerns about helicopter parents or those overly involved in students' lives inform institutional policies and practices, our findings suggest that campuses should aim to support family involvement in and support of FGCS experiences.

The recommendations we make here are based largely on our project; however, we recognize that there are already a number of promising interventions under way that offer important insights to higher education leaders. Certainly, some institutions are experimenting with requiring e-books or making 1:1 iPads available for low-income or FGCS. Our findings support these endeavors and suggest that when financial support is available for these technologies, FGCS benefit both directly and indirectly. For the FGCS in our study, having an iPad provided access to supportive apps and served as a status symbol at a time when students felt like outsiders on campus.

One promising intervention that warrants mention is the development of campus-specific apps that support students' transition to college. For example, Ball State University administrators created Ball State Achievements, an app designed to use gaming to introduce students to the college experience (Brown, 2015). The app, which was specifically targeted to students who were Pell Grant recipients, is being used to make sure FGCS gain information and skills necessary to be successful at Ball State. When students engage on campus (e.g., swiping into the library or the gym, attending a campus lecture, or joining a student organization), they earn points that can be used to purchase items in the campus bookstore. Ball State University administrators, who were initially drawn to the idea based on a Weight Watchers app, felt the type of engagement mattered less than the peer experience that accompanied it (Raths, 2015). Given that the university is finding that users of the app are graduating at higher rates, we imagine that campuses might want to build similar apps (Brown, 2015). These types of interventions take advantage of the fact that students overwhelmingly use smartphones on campus and reinforce the belief that becoming involved and engaged improves student outcomes (Raths, 2015).

A similar strategy to app development is the use of nudging technologies. This technology draws on behavioral science as a way to promote student success in college. The rationale behind these interventions is that by drawing on research from behavioral science, campus administrators can nudge students to complete important tasks and meet deadlines associated with successfully pursuing higher education. Nudging might take the form of emails or text messaging, but the goal is the same: to communicate something to students that will influence how they will make decisions or act at a given time or around a given event. In their report, *Nudging for Success*, ideas42 presented 16 programs that use behavioral science to support the postsecondary experience (Supiano, 2016). Other research that has explored the use of nudges to affect college-going behavior is also premised on behavioral science principles (Castleman & Page, 2015, 2016)

Perhaps the most intriguing opportunities from this study have to do with its scalability. Since Web 2.0 technologies are ubiquitous, they are relatively cost neutral to utilize with students. Additionally, the findings of the study call into question long-held assumptions about the role of integration into the campus environment as integral for student success. Although this may have been true when the only means of connecting with home communities was by going home or calling home from the pay phone bank in the dorm hallway, Skype, Facebook, and Instagram have changed the way individuals communicate. Consequently, students can use these technologies to connect with home communities for support essential for their success. Moreover, such technologies require less capital transfer to employ for resource and knowledge acquisition. A student may be reticent to raise her hand in class and ask a question because she does not want to make any knowledge deficiencies known, but our study indicates she may be far more likely to ask the same question of her FGCS peers through Facebook or another social media platform. Technology seems to mitigate the capital required to engage in such activities. Based on the findings presented in this book, we hope campus administrators consider some of these opportunities to leverage such technology to encourage student success.

ADVANCING THE SCHOLARSHIP

This book contributes to a body of scholarship on FGCS and campus-based technologies. The findings shared in this book describe the unique ways in which FGCS interact with social media and Web 2.0 technologies. In addition to new programs and practices that support students' natural use of technology, more research is needed to gain a deeper understanding of the specific mechanisms at play for FGCS using technology on campuses today.

The FGCS in our study were largely underrepresented students of color on a predominately White campus. In this way, students' social networks crossed race and class boundaries, yet this was not explored in this project. We anticipate that a deeper examination of how students develop or manage networks across such boundaries would be beneficial to the field. Relatedly, when reflecting on their engagement on campus, our FGCS described instances of racism and discrimination that take place on social media. New scholarship on the impact of online racism as it relates to student experiences, transitions, and sense of belonging is sorely needed. FGCS described difficult experiences that hindered their sense of belonging on campus. Yet, many of the FGCS in our study described feeling connected on campus. This raises questions about whether students differentiate their sense of belonging and connections to a given campus from affinity or connection to a particular group or counterculture group.

Without question, FGCS, like all students, are deeply engaged in using technology and social media on campus. Yet, because technology changes so quickly, it is difficult for campus administrators to stay current. If institutions are to utilize new forms of technology, they need up-to-date, relevant research on how new forms of communication using technology is taking place on college campuses. FGCS described the growing usage of Snapchat and specifically, how seeing images of peers around campus gives them a sense of connection to a place.

Mentoring on college campuses is growing in popularity, especially in those that operate in service to supporting students who may be struggling during the transition to college. Yet, our findings suggest that technology might be leveraged to support mentoring relationships and networks. Future research might examine the changing face of mentoring on college campuses.

FGCS strong ties with family and peers from home, and the benefits accrued from those relationships, raises questions about how FGCS manage those relationships while developing new ties on campus. Future research might consider examining these strong ties more deeply to understand how campuses can support these relationships.

Finally, the EOP in our study provided an important context to examine FGCS use of technology to support the transition to college and classroom learning. However, programs like the EOP, which operate as bridge programs, have not been widely researched, despite their presence on campuses across the country. More specific to our project, however, the staff and faculty in these programs operated first as weak ties, and then as quasi-strong ties, thereby acting as an important source of campus capital. Given their unique role and potential for supporting FGCS, we need to better understand what these programs do, how they work, and, importantly, their collective benefit to universities.

Above all, throughout our time with FGCS at Millbrook, it was clear that Web 2.0 technology impacted our students' efforts to progress through the social and academic transitions of their first years as college students. Social media, iPads, smartphones, image messaging, and multimedia mobile applications composed the landscape of their campus experiences. Whether and how higher education embraces and lays claim to the benefits of these technologies will determine whether and how institutions can truly make technology work for first-generation college students.

Acknowledgments

We undertook this project for professional and personal reasons. All three of us have worked extensively with first-generation college students (FGCS) and know from experience how challenging the transition to college can be. We have worked alongside faculty and staff who deeply care about supporting FGCS, yet are unsure about how to connect with them through technology in meaningful ways. Those experiences shaped our thinking about the need to leverage social media and Web 2.0 technologies in new ways and in ways that reflect a genuine understanding of FGCS experiences in higher education. Personally, two of us are first-generation college students and ourselves navigated the tricky seas of higher education. Collectively, we share a commitment to raising awareness about FGCS unique experiences and sharing results that have clear relevance for those working in the field of higher education. We hope our presentation of FGCS at Millbrook University reflects the genuine interest we have in their success.

This project benefited from the generosity and wisdom of many people. Foremost, we are deeply grateful to the Educational Opportunity Program (EOP) students who participated in our project and shared their ideas with our team. We learned a lot from these aspiring college graduates and applaud their commitment to earning a college degree in spite of the challenges they face. Likewise, a huge round of thanks is owed to the EOP staff and faculty at Millbrook University, all of whom were supportive and instructive to our project and learning. We had the incredible good fortune to have a terrific group of graduate students working on this project. Together, Kevin Gin, Adam Gismondi, Derek Hottell, Michele Brown Kerrigan, Sara Knight, Jonathan Lewis, Adam McCready, and Scott Radimer provided valuable contributions to multiple aspects of this project over the past five years. We are also deeply grateful to Millbrook University for its support for our project. Finally, we would like to thank Rutgers University Press for their interest in and support for this book.

Appendix: Research Methods

To develop a deeper understanding of how first-generation college students (FGCS) use technology, and how that is related to social and academic engagement, the research team utilized a case study design with a specific population of students in a specific programmatic setting operating for a set period of time (Yin, 2009). The program was a summer bridge program at an urban (Northeast United States), private, Predominantly White Institute (PWI) with approximately 9,000 undergraduate students. Pseudonyms were used for the names of the program, the institution, and all the individuals in the research study to ensure their privacy and encourage individuals to feel more comfortable sharing their opinions.

Selection of the site was based on prior relationships with the program staff at the site, program development of a Facebook page, and the provision of institutional funding to support the research. Although students came into our study as a part of the Educational Opportunity Program (EOP), it is the actual student participants who are our unit of analysis. During the summer of the 2012–2013 school year, there were 43 first-year students in the sample. During the 2013–2014 year, 40 new first-year students participated in the study. From 2014 to 2016, we followed 18 students who were part of one of our first two cohorts during their time at Millbrook. Students from the first cohort graduated in May 2016, and students from the second cohort were seniors at the start of the 2016–2017 academic year. Although it is not a requirement for selection, most of these students were low-socioeconomic status (SES) first-generation and more than 85% were students of color (SOC), as noted in the Introduction.

Students who consented to participate in the research project were provided with new iPads for their personal usage over the course of the summer and through their first year. Students were encouraged to use their iPads during their EOP English and mathematics classes, and English and mathematics instructors were also given iPads to utilize for the summer when teaching their classes. The iPads were preloaded with academic and social media apps, which the research team in consultation with the instructors believed would be beneficial academically and socially. The Office of Multicultural Affairs (OMA), which coordinates the program, also created a Facebook group for these students, with the assistance of the research team, to promote social networking within the group.

The research team, which included three faculty members and eight graduate research assistants over the time of the study, utilized a transformative, convergent mixed-method approach to draw on multiple sources of data to attain a fuller and more nuanced understanding of student engagement related to technology (Creswell & Plano Clark, 2011; Mertens, 2003). Members of the research team (one Latina, three White women, one Asian man, one biracial woman, and four White men) conducted all of the focus groups, interviews, and observations. Four interviewers were FGCS. Such an approach provided opportunities for continuous reflection throughout the research process, closer collaboration with the participants of the study, and further probing of incongruent findings (Creswell & Plano Clark, 2011). The uses of such approaches were vital to achieve the aims of the project and more comprehensively explored social media and tablet technology as mechanisms to promote academic and social engagement for historically marginalized students, as they transition into their first year of college.

Quantitative Data Collection

To partially answer how these FGCS used technology, and how they engaged academically and socially over their first year of college, quantitative data were collected. The research team collected the data over the course of the summer during the EOP and throughout the academic year from multiple sources. In addition to completing the Cooperative Institutional Research Program (CIRP) Freshman Survey as part of the university's orientation program, students also completed an online pretest inventory in June prior to the start of the summer program, consisting of approximately 30 multiple-choice and Likert item questions about their anticipated sense of belonging, sense of self, intended participation and engagement, social networking usage, and iPad familiarity and usage. Students would rank the degree to which each statement was "true for them." Students also completed a post-test survey at the completion of the summer program in August as well as a follow-up survey near the end of the spring semester in March assessing their first-year experiences.

Qualitative Data Collection

To help answer what role technology plays in student lives, and how it is related to social and academic engagement, a qualitative approach was also employed. The qualitative data were collected through classroom observations during the EOP; semi-structured individual interviews with students, faculty, and staff; and focus groups with EOP students, mentors, and counselors. Semi-structured interviews were utilized to afford researchers and participants the opportunity to explore the phenomena, as they were presented, while also providing focus for the interviews. Qualitative data was collected through 67 individual student interviews and 8 student focus groups over 4 years. Qualitative data were collected in early fall each year in the form of focus groups, and each year in the spring in the form of semi-structured individual interviews. Interviews lasted between 45 and 60 minutes and were recorded and transcribed. Interviews were designed to examine student perceptions of social media participation, specifically Facebook, both institutionally sponsored and students' personal accounts during the past year. Questions included: In what ways do you use social media to connect with people on and off campus? How do you learn how to do things on campus? Do you think any of your social media could be used to promote academic success?

The research team met regularly with EOP staff members to talk about social networking development, implementation, and data collection processes. These professionals were a part of the planning process for the project, and were also active participants. Some faculty and staff received iPads for use in the classroom, and members of the research team provided some professional development about how they could use iPad technology and social media in the classroom. At the end of each summer, EOP faculty and staff were interviewed. Four English faculty were interviewed at the end of the first and second summers, and four mathematics faculty (who received iPads only for the second year of the project) were interviewed during the second year. Four OMA staff were also interviewed each year. These semi-structured interviews were designed to examine faculty and staff perceptions of social media and technology (specifically iPad) use during the program. Questions included: In what ways do you use social media to connect with people on and off campus? How did you use social media or iPads in the classroom over the summer? How do you see students using technology in the EOP program? Do you think any of your social media could be used to promote academic success?

Members of the research team also conducted 36 class observations during the EOP. Over the first 2 years, 22 English classes were observed, and in the second year 14 mathematics classes were observed. Researchers were looking for ways in which technology was utilized in the classroom by both students and faculty members.

Analysis

The quantitative data were entered into SPSS for analysis. Descriptive statistics were calculated for all demographic variables, and then t-tests and chi-square analyses were conducted with the data.

The qualitative data were transcribed and then coded for themes using HyperResearch software. All interviews were transcribed verbatim and screen shots were taken of all Facebook activity on the EOP Facebook page. Utilizing a constant comparative method of data analysis (Charmaz, 2006), two members of the research team generated codes for each transcript and compared for consistency. Codes from the transcripts and the screen shots were then examined for relationships and patterns to create categories (Charmaz, 2006). The research team then developed themes that emerged within the case (Yin, 2009). Throughout the data analysis process, members of the research team continuously returned to the transcripts to confirm that our interpretations were true to students' words and experiences.

The research team utilized several strategies to ensure trustworthiness. The themes that emerged from the interviews were triangulated with information from the classroom observations, interviews with instructors and staff in the EOP, and the quantitative survey data. With several members of the research team engaged in analysis and numerous phases of data collection, the team had several opportunities to triangulate findings; to reflect on the intervention; and to communicate with the EOP students, staff, and faculty. Cross-check codes were also utilized to check for inter-coder reliability (Guest, MacQueen, & Namey, 2012). Additionally, members of the research team maintained thorough field notes during all points of data collection and conducted weekly team meetings to ensure consistency.

References

Acar, A. (2013). Attitudes toward blended learning and social media use for academic purposes: An exploratory study. *Journal of e-Learning and Knowledge Society*, *9*(3), 107–126.

Advisory Committee for Student Financial Assistance. (2001). *Access denied: Restoring the nation's commitment to equal educational opportunity*. Washington, DC: Author.

Alexander, B. (2006). Web 2.0: A new wave of innovation for teaching and learning? *EDUCAUSE Review*, *41*(2), 33–44.

Altbach, P. G., & Cohen, R. (1990). American student activism: The post-sixties transformation. *The Journal of Higher Education*, *61*, 32–49.

American Press Institute. (2014). Social and demographic differences in news habits and attitudes. Retrieved from http://www.americanpressinstitute.org/publications/reports/survey-research/social-demographic-differences-news-habits-attitudes/.

Anderson, M. (2015). Technology device ownership: 2015. Pew Research Center Report. Retrieved from http://www.pewinternet.org/2015/10/29/technology-device-ownership-2015.

Angst, C. M., & Malinowski, E. (2010). *Findings from the e-reader project, Phase 1: Use of iPads in MGMT40700, Project Management*. University of Notre Dame Working Paper Series. Retrieved from https://www3.nd.edu/~cangst/NotreDame_iPad_Report_01–06–11.pdf.

Anheier, H., & Kendall, J. (2002). Interpersonal trust and voluntary associations: Examining three approaches. *British Journal of Sociology*, *53*, 343–362. doi:10.1080/0007131022000000545.

Aries, E., & Seider, M. (2007). The role of social class in the formation of identity: A study of public and elite private college students. *Journal of Social Psychology*, *147*, 137–158. doi:10.3200/SOCP.147.2.137–157.

Armstrong, E., & Hamilton, L. (2013). *Paying for the party: How college maintains inequality*. Cambridge, MA: Harvard University Press.

Aud, S., Hussar, W., Johnson, F., Kena, G., Roth, E., Manning, E., Wang, X., & . . . Zhang, J. (2012). *The condition of education 2012*. NCES Publication No. 2012–045. Retrieved from http://nces.ed.gov/pubsearch.

Bailyn, B., Fleming, D., Handlin, O., & Thernstrom, S. (1995). *Glimpses of the Harvard past*. Cambridge, MA: Harvard University Press.

Balemian, K., & Feng, J. (2013, July). *First generation students: College aspirations, preparedness and challenges*. Symposium conducted at the College Board AP Annual Conference, Las Vegas, NV. Retrieved from https://research.collegeboard.org/sites/default/files/publications/2013/8/presentation-apac-2013-first-generation-college-aspirations-preparedness-challenges.pdf.

Baum, S., Ma, J., & Payea, K. (2013). *Education pays 2013: The benefits for individuals and society*. The College Board Trends in Higher Education Series. Retrieved from https://trends.collegeboard.org/sites/default/files/education-pays-2013-full-report-022714.pdf.

Bell, A. D., Rowan-Kenyon, H. T., & Perna, L. W. (2009). College knowledge of 9th and 11th grade students: Variation by school and state context. *The Journal of Higher Education*, *80*(6), 663–685.

Benson, V., Filippaios, F., & Morgan, S. (2010). Applications of social networking in students' life cycle. In C. Wankel (Ed.), *Cutting-edge social media approaches to business education: Teaching with LinkedIn, Facebook, Twitter, Second Life, and blogs* (pp. 73–93). Charlotte, NC: Information Age Publishing.

Bettinger, E. P., & Long, B. T. (2005). Remediation at the community college: Student participation and outcomes. In C. A. Kozeracki (Ed.), *New directions for community colleges* (pp. 17–26). San Francisco, CA: Jossey-Bass.

Bonilla, Y., & Rosa, J. (2015). #Ferguson: Digital protest, hashtag ethnography, and the racial politics of social media in the United States. *American Ethnologist*, *42*(1), 4–17.

Bourdieu, P. (1986). The forms of capital (R. Nice, Trans.). In J. F. Richardson (Ed.), *Handbook of theory and research for sociology and education* (pp. 241–258). New York: Greenwood. Reprinted from R. Kreckel (Ed.), *Soziale Ungleichheiten: Soziale Welt, Sonderheft 2* (pp. 183–198). Goettingen: Otto Schartz & Co., 1983.

Bourdieu, P., & Passeron, J. (1977). *Reproduction in education, society and culture*. London: Sage.

Bowen, W. G., Kurzweil, M. A., & Tobin, E. (2005). *Equity and excellence in American higher education*. Charlottesville: University of Virginia Press.

Brand, J., Kinash, S., Mathew, T., & Kordyban, R. (2011). iWant does not equal iWill: Correlates of mobile learning with iPads, e-textbooks, Blackboard Mobile Learn and a blended learning experience. In G. Williams, P. Statham, N. Brown, & B. Cleland (Eds.), *Changing demands, changing directions: Proceedings of Ascilite Hobart 2011* (pp. 168–178). Hobart: University of Tasmania Press.

Brooks, B., Hogan, B., Ellison, N., Lampe, C., & Vitak, J. (2014). Assessing structural correlates to social capital in Facebook ego networks. *Social Networks, 38*, 1–15.

Brooks, B., Welser, H. T., Hogan, B., & Titsworth, S. (2011). Socioeconomic status updates: Family SES and emergent social capital in college student Facebook networks. *Information, Communication & Society, 14*, 529–549. doi:10.1080/1369118X.2011.562221.

Brown II, M. C., & Dancy II, T. E. (2010). Predominantly White institutions. In K. Lomotey (Ed.), *Encyclopedia of African American education* (pp. 524–527). Thousand Oaks, CA: Sage. Retrieved from http://dx.doi.org/10.4135/9781412971966.n193.

Brown, M. G., Wohn, D. Y., & Ellison, N. (2016). Without a map: College access and the online practices of youth from low-income communities. *Computers & Education, 92*, 104–116. doi:10.1016/j.compedu.2015.10.001.

Brown, S. (2015, August 24). How an app helps low-income students by turning college life into a game. *The Chronicle of Higher Education*. Retrieved from http://chronicle.com/blogs/wiredcampus/how-an-app-helps-low-income-students-by-turning-college-life-into-a-game/57229?cid=megamenu.

Burke, M., Kraut, R., & Marlow, C. (2011). Social capital on Facebook: Differentiating uses and users. *Proceedings of the SIGCHI Conference on Human Factors in Computing Systems* (pp. 571–580). New York: ACM. doi:10.1145/1978942.1979023.

Cao, Y., Aijan, H., & Hong, P. (2013). Using social media applications for educational outcomes in college teaching: A structural equation analysis. *British Journal of Educational Technology, 44*(4), 581–593. doi:10.1111/bjet.12066.

Carolan, B., & Natriello, G. (2006). *Strong ties, weak ties: Relational dimensions of learning settings*. EdLab Technical Report. Retrieved from http://edlab.tc.columbia.edu/files/EdLab_Strongties.pdf.

Castleman, B. L., & Page, L. C. (2015). Summer nudging: Can personalized text messages and peer mentor outreach increase college going among low-income high school graduates? *Journal of Economic Behavior & Organization, 115*, 144–160.

Castleman, B. L., & Page, L. C. (2016). Freshman year financial aid nudges: An experiment to increase FAFSA renewal and college persistence. *Journal of Human Resources, 51*(2), 389–415.

Center for Urban Education. (2016). *CUE's impact on equity gaps*. Retrieved from https://cue.usc.edu/equity/impact/.

Chaffey, D. (2015). Insights from KPCB US and global internet trends 2015 report. Retrieved from http://www.smartinsights.com/internet-marketing statistics/insights-from-kpcb-us-and-global-internet-trends-2015-report/.

Charles A. Dana Center, Complete College America, Education Commission of the States, & Jobs for the Future. (2012). *Core principles for transforming remedial education: A joint statement*. Retrieved from http://www.jff.org/publications/core-principles-transforming-remedial-education-joint-statement.

Charmaz, K. (2006). *Constructing grounded theory: A practical guide through qualitative analysis*. Thousand Oaks, CA: Sage.

Chen, P.S.D., Lambert, A. D., & Guidry, K. R. (2010). Engaging online learners: The impact of web-based learning technology on college student engagement. *Computers & Education, 54*, 1222–1232.

Chen, X. (2005). First generation students in postsecondary education: A look at their college transcripts. NCES 2005–171. Retrieved from http://nces.ed.gov/pubs2005/2005171.pdf.

Cheng, H., Hitter, T. L., Adams, E. M., & Williams, C. (2016). Minority stress and depressive symptoms: Familism, ethnic identity, and gender as moderators. *The Counseling Psychologist, 44*, 841–870. Retrieved from https://doi.org/10.1177/0011000016660377.

Cheung, C. M., Chiu, P. Y., & Lee, M. K. (2011). Online social networks: Why do students use Facebook? *Computers in Human Behavior, 27*(4), 1337–1343.

Coleman, J. S. (1988). Social capital in the creation of human capital. *American Journal of Sociology, 94*(Supplement), 95–120.

comScore. (2015). *2015 U.S. digital future in focus*. comScore White Paper. Retrieved from http://www.comscore.com/Insights/Presentations-and-Whitepapers/2015/2015-US-Digital-Future-in-Focus.

Covarrubias, R., & Fryberg, S. A. (2015). Movin' on up (to college): First-generation college students' experiences with family achievement guilt. *Cultural Diversity and Ethnic Minority Psychology, 21*(3), 420–429. Retrieved from http://dx.doi.org/10.1037/a0037844.

Cox, R. (2011). *The college fear factor: How students and professors misunderstand one another*. Cambridge, MA: Harvard University Press.

Creswell, J. W., & Plano Clark, V. (2011). *Designing and conducting mixed methods research*. Los Angeles, CA: Sage.

Cushman, K. (2006). *First in the family: Advice about college from first-generation students*. Providence, RI: Next Generation Press.

Dabbagh, N., & Kitsantas, A. (2011). Personal Learning Environments, social media, and self-regulated learning: A natural formula for connecting formal and informal learning. *Internet and Higher Education*. doi:10.1016/j.iheduc.2011.06.002.

Dabbagh, N., & Reo, R. (2011). Impact of Web 2.0 on higher education. In D. W. Surry, T. Stefurak, & R. Gray (Eds.), *Technology integration in higher education: Social and organizational aspects* (pp. 174–187). Hershey, PA: IGI Global.

Dahlstrom, E. (2012). *ECAR Study of Students and Information Technology*. Research report. Louisville, CO: ECAR. Retrieved from http://www.educause.edu/ecar.

Dahlstrom, E., with Brooks, D. C., Grajek, S., & Reeves, J. (2015). *ECAR Study of Students and Information Technology*. Research report. Louisville, CO: ECAR.

Dahlstrom, E., Walker, J. D., & Dziuban, C. (2013). *ECAR Study of Students and Information Technology*. Research report. Louisville, CO: ECAR. Retrieved from http://net.educause.edu/ir/library/pdf/ERS1302/ERS1302.pdf.

Daniel Tatum, B. (1997). *"Why are all the Black kids sitting together in the cafeteria?" and other conversations about the development of racial identity.* New York: Basic Books.

Davis, J. (2010). *The first generation student experience: Implications for campus practice, and strategies for improving persistence and success.* Sterling, VA: Stylus Publishing.

DeAndrea, D. C., Ellison, N. B., LaRose, R., Steinfield, C., & Fiore, A. (2011). Serious social media: On the use of social media for improving students' adjustment to college. *Internet and Higher Education, 15*, 15–23. doi:10.1016/j.iheduc.2011.05.009.

Delello, J. A., McWhorter, R. R., & Camp, K. M. (2015). Using social media as a tool for learning: A multi-disciplinary study. *International Journal on E-Learning, 14*(2), 163–180.

Dennis, J. M., Phinney, J. S., & Chuateco, L. I. (2005). The role of motivation, parental support, and peer support in the academic success of ethnic minority first-generation college students. *Journal of College Student Development, 46*, 223–236.

Dewey, J. (1915). *The school and society.* Chicago, IL: University of Chicago Press.

DiMaggio, P. (1982). Cultural capital and school success. *American Sociological Review, 47*, 189–201.

Dogtiev, A. (2015). App usage statistics: 2015 roundup. Retrieved from http://www.businessofapps.com/app-usage-statistics-2015/.

Dryer, K. (2015). Mobile internet usage skyrockets in past 4 years to overtake desktop as most used digital platform. Retrieved from https://www.comscore.com/Insights/Blog/Mobile-Internet-Usage-Skyrockets-in-Past-4-Years-to-Overtake-Desktop-as-Most-Used-Digital-Platform.

Duggan, M., & Brenner, J. (2013). *The demographics of social media users—2012.* Washington, DC: Pew Internet Project. Retrieved from http://www.pewinternet.org/2013/02/14/the-demographics-of-social-media-users-2012/.

Duggan, M., Ellison, N. B., Lampe, C., Lenhart, A., & Madden, M. (2015, January 9). *Social Media Update 2014.* Retrieved from http://www.pewinternet.org/files/2015/01/PI_SocialMediaUpdate20144.pdf.

Duggan, M., & Smith, A. (2013). Demographics of key social networking platforms. *Social Media Update 2013.* Retrieved from http://www

.pewinternet.org/2013/12/30/demographics-of-key-social-networking-platforms/.

Easley, D., & Kleinberg, J. (2010). *Networks, crowds, and markets: Reasoning about a highly connected world.* Cambridge: Cambridge University Press.

Ellison, N. B., Lampe, C., & Steinfield, C. (2009). Social network sites and society: Current trends and future possibilities. *Interactions, 16*(1), 6–9.

Ellison, N. B., Steinfield, C., & Lampe, C. (2007). The benefits of Facebook "friends": Social capital and college students' use of online social network sites. *Journal of Computer-Mediated Communication, 12*(4), article 1. Retrieved from http://jcmc.indiana.edu/vol12/issue4/ellison.html.

Ellison, N. B., Steinfield, C., & Lampe, C. (2011). Connection strategies: Social capital implications of Facebook-enabled communication practices. *New Media & Society, 13*(6), 873–892. doi:10.1177/1461444810385389.

Ellison, N. B., Vitak, J., Gray, R., & Lampe, C. (2014). Cultivating social resources on social network sites: Facebook relationship maintenance behaviors and their role in social capital processes. *Journal of Computer-Mediated Communication, 19*(4), 855–870.

Ellison, N. B., & Wu, Y. (2008). Blogging in the classroom: A preliminary exploration of student attitudes and impact on comprehension. *Journal of Educational Multimedia and Hypermedia, 17*(1), 99–122.

Engle, J., & Tinto, V. (2008). *Moving beyond access: College success for low-income, first-generation students.* Washington, DC: The Pell Institute.

Ferrer, F., Belvís, E., & Pàmies, J. (2011). Tablet PCs, academic results and educational inequalities. *Computers & Education, 56*(1), 280–288.

Fleming, J. (1985). *Blacks in college. A comparative study of students' success in Black and in White institutions.* San Francisco, CA: Jossey-Bass.

Florini, S. (2014). Tweets, tweeps, and signifyin': Communication and cultural performance on "Black Twitter." *Television & New Media, 15*(3), 223–237.

Gil de Zúñiga, H., Jung, N., & Valenzuela, S. (2012). Social media use for news and individuals' social capital, civic engagement and political participation. *Journal of Computer–Mediated Communication, 17*(3), 319–336.

Gilbert, E., & Karahalios, K. (2009). Predicting tie strength with social media. *Proceedings of the 27th international conference on human factors in computing systems* (pp. 211–220). New York: ACM. Retrieved from http://dl.acm.org/citation.cfm?id=1518736.

Gin, K. J., Martínez Alemán, A. M., Knight, S., Radimer, S., Lewis, J., & Rowan-Kenyon, H. T. (2016). Democratic education online: Combating racialized aggressions on social media. *Change: The Magazine of Higher Learning, 48*(3), 28–35. doi:10.1080/00091383.2016.1170531.

Golder, S., Wilkerson, D., & Huberman, B. (2005). Rhythms of social interaction: Messaging within a massive online network. Retrieved from http://www.hpl.hp.com/research/idl/papers/facebook/facebook.pdf.

Granovetter, M. (1973). The strength of weak ties. *American Journal of Sociology, 78*(6), 1360–1380.

Granovetter, M. (1983). The strength of weak ties: A network theory revisited. *Sociological Theory, 1*, 201–233.

Gray, R., Vitak, J., Easton, E. W., & Ellison, N. B. (2013). Examining social adjustment to college in the age of social media: Factors influencing successful transitions and persistence. *Computers & Education, 67*, 193–207.

Greenhow, C., & Gleason, B. (2012). Twitteracy: Tweeting as a new literacy practice. *Educational Forum, 76*, 464–478.

Greenwood, S., Perrin, A., & Duggan, M. (2016). *Social Media Update 2016*. Retrieved from http://www.pewinternet.org/2016/11/11/social-media-update-2016/.

Guest, G., MacQueen, K., and Namey, E. E. (2012). *Applied thematic analysis*. Los Angeles, CA: Sage.

Habermas, J. (1981). *The theory of communicative action, Volume 1: Reason and the rationalization of society*. Boston, MA: Beacon Press.

Habermas, J. (1987). *The theory of communicative action, Volume 2: Lifeworld and system: A critique of functionalist reason*. Boston, MA: Beacon Press.

Hall Jr., O. P., & Smith, D. M. (2011). Assessing the role of mobile learning systems in graduate management education. *Hybrid learning* (pp. 279–288). Heidelberg: Springer Berlin.

Hamid, S., Waycott, J., Kurnia, S., & Chang, S. (2015). Understanding students' perceptions of the benefits of online social networking use for teaching and learning. *The Internet and Higher Education, 26*, 1–9. doi:10.1016/j.iheduc.2015.02.004.

Hampton, K., Goulet, L. S., Marlow, C., & Rainie, L. (2012, February 3). *Part 3: The structure of friendship*. Retrieved from http://www.pewinternet.org/2012/02/03/part-3-the-structure-of-friendship/.

Hansen, M. T. (1999). The search-transfer problem: The role of weak ties in sharing knowledge across organization subunits. *Administrative Science Quarterly*, *44*(1), 82–111.

Harackiewicz, J. M., Canning, E. A., Tibbetts, Y., Giffen, C. J., Blair, S. S., Rouse, D. I., & Hyde, J. S. (2013, November 4). Closing the social class achievement gap for first-generation students in undergraduate biology. *Journal of Educational Psychology*. Advance onlinepublication. doi:10.1037/a0034679.

Harker, R. (1990). Education and cultural capital. In R. Harker, C. Mahar, & C. Wilkes (Eds.), *An introduction to the work of Pierre Bourdieu: The practice of theory*. London: Macmillan Press.

Hayes, L. L. (1997). Support from family and institution crucial to success of first-generation college students. *Counseling Today*, *40*(2), 1–4.

Heiberger, G., & Harper, R. (2008). Have you facebooked Astin lately? Using technology to increase student involvement. *New Directions For Student Services*, *124*, 19–35. doi:10.1002/ss.293.

Heinrich, P. (2012). The iPad as a tool for education: A study of the introduction of iPads in Longfield Academy, Kent. NAACE/ICT Association. Retrieved from http://www.emergingedtech.com/2012/07/study-finds-benefits-in-use-of-ipad-as-educational-tool/.

Hofer, M., & Aubert, V. (2013). Perceived bridging and bonding social capital on Twitter: Differentiating between followers and followees. *Computers in Human Behavior*, *29*, 2134–2142. doi:10.1016/J.CHB.2013.04.038.

Horgen, S. A., & Olsen, T. O. (2014). Effects of social media on building and using personal learning networks. In L. Gómez Chova, A. López Martínez, & I. Candel Torres (Eds.), *INTED2014 Proceedings: 8th International Technology, Education and Development Conference* (pp. 3553–3560). Valencia, Spain: IATED Academy.

Horkheimer, M. (1982). *Critical theory*. New York: Seabury Press.

Horvat, E. M. (2000). Understanding equity and access in higher education: The potential contribution of Pierre Bourdieu. In J. C. Smart & W. Tierney (Eds.), *The higher education handbook of theory and research* (Vol. 16, pp. 195–238). Hingham, MA: Kluwer Academic Publishers.

How-to-Geek. (2016). Tablets aren't killing laptops, but smartphones are killing tablets. Retrieved from http://www.howtogeek.com/199483/tablets-arent-killing-laptops-but-smartphones-are-killing-tablets/.

Hurtado, S., & Carter, D. F. (1997). Effects of college transition and perceptions of the campus racial climate on Latino college students' sense of belonging. *Sociology of Education, 70*, 324–345. Retrieved from http://dx.doi.org/10.2307/2673270.

Inman, W. I., & Mayes, L. (1999). The importance of being first generation community college students. *Community College Review, 26*(4), 3–22.

Institute for Higher Education Policy. (2011). *Maximizing the college choice process to increase fit & match for underserved students*. Washington, DC: Author.

Institute for Higher Education Policy. (2012). *Supporting first-generation college students through classroom-based practices*. Washington, DC: Author.

Jehangir, R. R. (2010a). *Higher education and first-generation students: Cultivating community, voice, and place for the new majority*. New York: Palgrave Macmillan.

Jehangir, R. R. (2010b). Stories as knowledge: Bringing the lived experience of first-generation college students into the academy. *Urban Education, 45*, 533–553.

Johnson, D., Soldner, M., Leonard, J. B., Alvarez, P., Inkelas, K. K., Rowan-Kenyon, H. T., & Longerbeam, S. (2007). Examining sense of belonging among first-year undergraduates from different racial/ethnic groups. *Journal of College Student Development, 48*, 525–542.

Junco, R. (2011). The relationship between frequency of Facebook use, participation in Facebook activities, and student engagement. *Computers and Education, 58*, 162–171.

Junco, R. (2013). Comparing actual and self-reported measures of Facebook use. *Computers in Human Behavior, 29*(3), 626–631. doi:10.1016/j.chb.2012.11.007.

Junco, R. (2014). *Engaging students through social media: Evidence-based practices for use in student affairs*. San Francisco, CA: Jossey-Bass.

Junco, R., & Cole–Avent, G. A. (2008). An introduction to technologies commonly used by college students. *New Directions for Student Services, 124*, 3–17.

Junco, R., Heiberger, G., & Loken, E. (2011). The effect of Twitter on college student engagement and grades. *Journal of Computer Assisted Learning, 27*(2), 119–132. doi:10.1111/j.1365–2729.2010.00387.x.

Kaufer, D., Gunawardena, A., Tan, A., & Cheek, A. (2011). Bringing social media to the writing classroom: Classroom salon. *Journal of Business and Technical Communication*. doi:10.1177/1050651911400703.

Kennedy, S., & Winkle-Wagner, R. (2014). Earning autonomy while maintaining family ties: Black women's reflections on the transition into college. *NASPA Journal About Women in Higher Education*, 7, 133–152.

Kezar, A., Walpole, M., & Perna, L. W. (2014). Engaging low-income students. In S. J. Quaye & S. R. Harper (Eds.), *Student engagement in higher education: Theoretical perspectives and practical approaches for diverse populations* (pp. 237–255). New York: Routledge.

Kim, Y., & Sax., L. (2009). Student-faculty interaction in research universities: Differences by student gender, race, social class, and first-generation status. *Research in Higher Education*, 50, 437–459.

Kinash, S., Brand, J., & Mathew, T. (2012). Challenging mobile learning discourse through research: Student perceptions of Blackboard Mobile Learn and iPads. *Australasian Journal of Educational Technology*, *28*(4), 639–655.

Krackhardt, D. (1992). The strength of strong ties: The importance of philos in organizations. In N. Nohria & R. Eccles (Eds.), *Networks and organizations: Structure, form, and action* (pp. 216–239). Boston, MA: Harvard Business School Press.

Kuh, G. D., Kinzie, J., Buckley, J. A., Bridges, B. K., & Hayek, J. C. (2007). *Piecing together the student success puzzle: Research, propositions, and recommendations*. ASHE Higher Education Report (Vol. 32, No. 5). San Francisco, CA: Jossey-Bass.

Kuh, G. D., Kinzie, J., Cruce, T., Shoup, R., & Gonyea, R. (2006). *Connecting the dots: Multifaceted analysis of the relationships between student engagement results from the NSSE and the institutional policies and conditions that foster student success*. Final report to Lumina Foundation for Education. Bloomington: Indiana University Center for Postsecondary Research.

Ladson-Billings, G., & Tate IV, W. F. (1995). Toward a critical race theory of education. *Teachers College Record*, *97*(1), 48–68.

Lamont, M., & Lareau, A. (1988). Cultural capital: Allusions, gaps, and glissandos in recent theoretical developments. *Sociological Theory*, *6*, 153–168.

Lenhart, A. (2015). *Teens, social media & technology overview 2015*. Retrieved from http://www.pewinternet.org/2015/04/09/teens-social-media-technology-2015/.

Lévi-Strauss, C. (1963). *Structural anthropology* (C. Jacobson & B. G. Schoepf, Trans.). New York: Basic Books.

Lin, N. (1999). Building a network theory of social capital. *Connections*, *22*, 28–51.

Lin, N. (2001a). Building a network theory of social capital. In N. Lin, K. Cook, & R. S. Burt (Eds.), *Social capital: Theory and research* (pp. 3–30). New York: Walter de Gruyter, Inc.

Lin, N. (2001b). *Social capital: A theory of social structure and action*. New York: Cambridge University Press.

Lin, N., Ensel, W. M., & Vaughn, J. C. (1981). Social resources and the strength of weak ties: Structural factors in occupational status attainment. *American Sociological Review*, *46*, 393–405.

Linder, C. (2008). A zero-sum game? The popular media and gender gap in higher education. *Journal of Student Affairs*, *17*, 48–54.

Liu, O. L., Bridgeman, B., & Alder, R. M. (2012). Measuring learning outcomes in higher education: Motivation matters. *Educational Researcher*, *41*(9), 352–362.

Lohfink, M., & Paulsen, M. B. (2005). Comparing the determinants of persistence for first-generation and continuing-generation students. *Journal of College Student Development*, *46*, 409–428.

London, H. B. (1989). Breaking away: A study of first-generation college students and their families. *American Journal of Education*, *97*, 144–170.

Manuguerra, M., & Petocz, P. (2011). Promoting student engagement by integrating new technology into tertiary education: The role of the iPad. *Asian Social Science*, *7*(11), 61–65. doi:10.5539/ass.v7n11p61.

Marmarelli, T., & Ringle, M. (2011). The Reed College iPad study. Retrieved from http://www.reed.edu/cis/about/ipad_pilot/Reed_ipad_report.pdf.

Martínez Alemán, A. M. (1999). ¿Qué culpa tengo yo? Performing identity and college teaching. *Educational Theory*, *49*(1), 37–51.

Martínez Alemán, A. M. (2000). Race talks: Undergraduate women of color and female friendship. *The Review of Higher Education*, *23*, 133–152.

Martínez Alemán, A. M. (2014, January/February). Social media go to college. *Change: The Magazine of Higher Learning*. doi:10.1080/00091383.2014.867203.

Martínez Alemán, A. M., & Wartman, K. L. (2009). *Online social networking on campus: Understanding what matters in student culture.* New York: Routledge/Francis Taylor Group.

McCorkle, C. S. (2012). First-generation, African American students' experiences of persisting at a predominantly White liberal arts college. Unpublished doctoral dissertation, Western Michigan University, Kalamazoo, MI. Retrieved from http://scholarworks.wmich.edu/dissertations/65/.

McLoughlin, C., & Lee, M.J.W. (2010). Personalised and self-regulated learning in the Web 2.0 era: International exemplars of innovative pedagogy using social software. *Australasian Journal of Educational Technology, 26*(1), 28–43.

McPherson, M., Smith-Lovin, L., & McCook, M. (2001). Birds of a feather: Homophily in social networks. *American Review of Sociology, 27*, 415–444.

Melguizo, T. (2008). Quality matters: Assessing the impact of attending more selective institutions on college completion rates of minorities. *Research in Higher Education, 49*, 214–236.

Mertens, D. M. (2003). Mixed methods and the politics of human research: The transformative emancipatory perspective. In A. Tashakori & C. Teddlie (Eds.), *Handbook of mixed methods in social and behavioral research* (pp. 135–164). Thousand Oaks, CA: Sage.

Moll, L. C., Amanti, C., Neff, D., & Gonzalez, N. (1992). Funds of knowledge for teaching: Using a qualitative approach to connect homes and classrooms. *Theory Into Practice, 31*(2), 132–141.

Moran, M., Seaman, J., & Tinti-Kane, H. (2011). Teaching, learning, and sharing: How today's higher education faculty use social media. Retrieved from http://www.babson.edu/Academics/Documents/babson-survey-research-group/teaching-learning-and-sharing.pdf.

Moran, M., Seaman, J., & Tinti-Kane, H. (2012). Blogs, wikis, podcasts, and Facebook: How today's higher education faculty use social media. Retrieved from http://www.onlinelearningsurvey.com/reports/blogswikispodcasts.pdf.

Morris, J., Reese, J., Beck, R., & Mattis, C. (2009). Facebook usage as a predictor of retention at a private 4-year institution. *Journal of College Student Retention, 11*(3), 311–322. Retrieved from http://doi.org/10.2190/CS.11.3.a.

Muñoz, C., & Towner, T. (2010). Social networks: Facebook's role in the advertising classroom. *Journal of Advertising Education, 14*(1), 20–27.

Muñoz, F. M., & Strotmeyer, K. C. (2010). Demystifying social media. *Journal of Student Affairs Research and Practice, 47*, 1–10. Retrieved from http://doi.org/10.2202/1949-6605.6132.

Murphy, G. D. (2011). Post-PC devices: A summary of early iPad technology adoption in tertiary environments. *E-Journal of Business Education & Scholarship of Teaching, 5*(1), 18–32.

Museus, S. D. (2014). The Culturally Engaging Campus Environments (CECE) Model: A new theory of college success among racially diverse student populations. *Higher Education: Handbook of Theory and Research, 29*, 189–227. doi10.1007/978-94-017-8005-6-5.

Museus, S. D. (2010). Delineating the ways that targeted support programs facilitate minority students' access to social networks and development of social capital in college. *Enrollment Management Journal, 4*, 10–41.

Nagele, C. (2005). Social networks research report. Wildbit, LLC. Retrieved from http://wildbit.com/wildbit-sn-report.pdf.

Newman, M. W., Lauterbach, D., Munson, S. A., Resnick, P., & Morris, M. E. (2011, March). It's not that I don't have problems, I'm just not putting them on Facebook: Challenges and opportunities in using online social networks for health. *Proceedings of the ACM 2011 conference on computer supported cooperative work* (pp. 341–350). New York: ACM.

Nguyen, L., Barton, S. M., & Nguyen, L. T. (2015). iPads in higher education hype and hope. *British Journal of Educational Technology, 46*(1), 190–203.

Nichols, L., & Islas, A. (2015). Pushing and pulling emerging adults through college: College generational status and the influence of parents and others in the first year. *Journal of Adolescent Research, 31*, 59–95. Retrieved from https://doi.org/10.1177/0743558415586255.

Nortcliffe, A., & Middleton, A. (2013). The innovative use of personal smart devices by students to support their learning. *Cutting-edge Technologies in Higher Education, 6*, 175–208.

Nuñez, A., & Cuccaro-Alamin, S. (1998). *First-generation students: Undergraduates whose parents never enrolled in postsecondary education*. Washington, DC: NCES.

Olson, A. (2011). First-generation college students' experiences with social class identity dissonance. Unpublished doctoral dissertation, University of Denver, Colorado. Retrieved from http://digitaldu.coalliance.org/fedora/repository/codu%3A63016/ETD_Olson_denver_0061D_10378.pdf-0/master.

Ong, A. D., Phinney, J. S., & Dennis, J. (2006). Competence under challenge: Exploring the protective influence of parental support and ethnic identity in Latino college students. *Journal of Adolescence*, *29*, 961–979. doi:10.1016/j.adolescence.2006.04.010.

O'Reilly, T. (2005, September 30). *What is Web 2.0: Design patterns and business models for the next generation of software*. Retrieved from http://oreilly.com/web2/archive/what-is-web-20.html.

Orbe, M. P. (2004). Negotiating multiple identities within multiple frames: An analysis of first-generation college students. *Communication Education*, *53*(2), 131–149.

Ostrove, J. M., & Long, S. M. (2007). Social class and belonging: Implications for college adjustment. *The Review of Higher Education*, *30*, 363–389.

Pascarella, E. T., Pierson, C. T., Wolniak, G., & Terenzini, P. (2004). First-generation college students: Additional evidence on college experiences and outcomes. *The Journal of Higher Education*, *75*, 249–284.

Pearson Foundation. (2011). The majority of students want their own tablet device. Retrieved from http://www.pearsoned.com/news/pearson-foundation-survey-students-tablets-favor-digital-textbooks/.

Pearson Foundation. (2012). *The second annual Pearson Foundation survey on students and tablets*. Retrieved from http://www.pearsonfoundation.org/downloads/PF_Tablet_Survey_Summary_2012.pdf.

Perrin, A. (2015, October 8). *Social media usage: 2005–2015*. Pew Research Center. Retrieved from http://www.pewinternet.org/2015/10/08/social-networking-usage-2005–2015/.

Pew Internet and American Life Project. (2013). *Teens and technology*. Retrieved from http://www.pewinternetorg/2013/05/2/part-1-teens-and-social-media-use/.

Pew Research Center. (2015). *The smartphone difference*. Retrieved from http://www.pewinternet.org/2015/04/01/us-smartphone-use-in-2015/.

Pike, G. R., & Kuh, G. D. (2005). A typology of student engagement for American colleges and universities. *Research in Higher Education*, *46*(2), 185–209. doi:10.1007/s 11162–004–1599–0.

Putnam, R. (2000). *Bowling alone: The collapse and revival of American community*. New York: Simon & Schuster.

Quaye, S., Griffin, K., & Museus, S. (2015). Engaging students of color. In S. J. Quaye & S. R. Harper (Eds.), *Student engagement in higher education: Theoretical perspectives and practical approaches for diverse populations* (pp. 15–35). New York: Routledge.

Raine, L., & Wellman, B. (2012). *Networked: The new social operating system*. Cambridge, MA: MIT Press.

Raths, D. (2015, August 19). Achieving student success through gamification. *Campus Technology*. Retrieved from https://campustechnology.com/articles/2015/08/19/achieving-student-success-through-non-academic-means.aspx.

Razfar, A. (2008). Developing technological literacy: A case study of technology integration in a Latina liberal arts college. *AACE Journal, 16*(3), 327–345.

Rendón, L. I., Jalomo, R. E., & Nora, A. (2000). Theoretical considerations in the study of minority student retention in higher education. In J. M. Braxton (Ed.), *Reworking the student departure puzzle* (pp. 127–156). Nashville, TN: Vanderbilt University Press.

Roblyer, M. D., McDaniel, M., Webb, M., Herman, J., & Witty, J. V. (2010). Findings on Facebook in higher education: A comparison of college faculty and student uses and perceptions of social networking sites. *Internet and Higher Education, 13*, 134–140.

Rodriguez, J. E. (2011). Social media use in higher education: Key areas to consider for educators. *Journal of Online Learning and Teaching, 7*(4).

Rosenbaum, J. E., Deil-Amen, R., & Person, A. E. (2006). *After admission: From college access to college success*. New York: Russell Sage.

Saenz, V., Hurtado, S., Barrera, D., Wolf, D., & Yeung, F. (2007). *First in my family: A profile of first-generation college students at four-year institutions since 1971*. Los Angeles, CA: Higher Education Research Institute.

Saguaro Seminar. (2000). *BetterTogether*. Retrieved from http://www.bettertogeterh.org/bt_report.pdf.

Sarcedo, G. L., Matias, C. E., Monotoya, R., & Nishi, N. (2015). Dirty dancing with race and class: Microaggressions toward first-generation and low income college students of color. *Journal of Critical Scholarship on Higher Education and Student Affairs*, *2*(1). Retrieved from http://ecommons.luc.edu/jcshesa/vol2/iss1/1.

Satoh, H., Akamatsu, S., Yoshida, M., Yamaguchi, T., Eguchi, F., & Higashioka, Y. (2015). Collaborative tablet PC the system for self-active awareness in a dormitory environment. In P. Zaphiris & A. Ioannou (Eds.), *Learning and Collaboration Technologies* (pp. 503–509). Basel: Springer International Publishing.

Saunders, M., & Serna, I. (2004). Making college happen: The college experiences of first-generation Latino students. *Journal of Hispanic Higher Education*, *3*(2), 146–163.

Savitz-Romer, M., & Bouffard, S. (2012). *Ready, willing and able: A developmental approach to college access and success*. Cambridge, MA: Harvard Education Press.

Savitz-Romer, M., Rowan-Kenyon, H. T., & Fancsali, C. (2015). Noncognitive skills for college and career success. *Change: The Magazine of Higher Learning*, *47*(5), 18–27.

Scarcella, R. (2003). *Academic English: A conceptual framework*. Linguistic Minority Research Institute Newsletter. University of California, Santa Barbara.

Schiffrin, H. H., Liss, M., Miles-McLean, H., Geary, K., Erchull, M. J., & Tashner, T. (2014). Helping or hovering? The effects of helicopter parenting on college students' well-being. *Journal of Child and Family Studies*, *23*, 548–557. doi:10.1007/s10826–013–9716–3.

Scott, J. (2012). *Social network analysis*. Thousand Oaks, CA: Sage.

Scott, J., & Carrington, P. J. (Eds.). (2011). *The SAGE handbook of social network analysis*. Thousand Oaks, CA: Sage.

Sharma, S. (2013). Black Twitter? Racial hashtags, networks, and contagion. *New Formations: A Journal of Culture/Theory/Politics*, *78*(1), 48–64.

Smith, J. E., & Tirumala, L. N. (2012). Twitter's effects on student learning and social presence perceptions. *Teaching Journalism and Mass Communications*, *2*(1), 21–31.

Spalter-Roth, R., Van Vooren, N., Kisielewski, M., & Senter, M. (2013). *Strong ties, weak ties, or no ties: What helped sociology majors find career-level jobs?* Washington, DC: American Sociological Association. Retrieved from http://www.asanet.org/documents/research/pdfs/Bach_Beyond5_Social_Capital.pdf.

St. John, E. P., Hu, S., & Fisher, A. S. (2010). *Breaking through the access barrier: Academic capital formation informing policy in higher education*. New York: Routledge.

Stanton-Salazar, R. D. (1997). A social capital framework for understanding the socialization of racial minority children and youths. *Harvard Educational Review*, *94*, 1–40.

Statista. (2016). *Number of iPad users in the United States from 2013 to 2020 (in millions)*. Retrieved from http://statista.com/statistics/208039/ipad-users-forecast-in-the-us/.

Steinfield, C., Ellison, N. B., & Lampe, C. (2008). Social capital, self-esteem, and use of online social network sites: A longitudinal analysis. *Journal of Applied Developmental Psychology*, *29*, 434–445.

Stephens, N. M., Brannon, T. N., Markus, H. R., & Nelson, J. E. (2015). Feeling at home in college: Fortifying school–relevant selves to reduce social class disparities in higher education. *Social Issues and Policy Review*, *9*(1), 1–24.

Stephens, N., Fryberg, S., Markus, H., Johnson, C., & Covarrubias, R. (2012). Unseen disadvantage: How American universities' focus on independence undermines the academic performance of first-generation college students. *Journal of Personality and Social Psychology*, *102*, 1178–1197. doi:10.1037/a0027143.

Stephens, N., Hamedani, M., & Destin, M. (2014). Closing the social-class achievement gap: A difference-education intervention improves first-generation students' academic performance and all students' college transition. *Psychological Science*, *25*, 943–953. doi:10.1177/0956797613518349.

Strayhorn, T. L. (2012). *College students' sense of belonging: A key to educational success for all students*. New York: Routledge.

Supiano, B. (2016, June 22). Fine tuning the nudges that help students get to and through college. *The Chronicle of Higher Education*. Retrieved from http://chronicle.com/article/Fine-Tuning-the-Nudges-That/236880.

Svendsen, L. P., & Mondahl, M. (2013). How social media enhanced learning platforms support students in taking responsibility for their own learning. *Journal of Applied Research in Higher Education*, *5*(2), 261–272.

Sy, S. R., Fong, K., Carter, R., Boehme, J., & Alpert, A. (2011). Parent support and stress among first generation and continuing-generation female students during the transition to college. *Journal of College Student Retention, 13*, 383–398.

Taylor, B. (2015). *5 ways the smartphone is conquering the tablet*. Retrieved from http://www.pcworld.com/article/2889275/5-ways-the-smartphone-is-conquering-the-tablet.htlml.

Taylor, M. (2010). Teaching generation NeXt: A pedagogy for today's learners. *2010 Higher Learning Commission Collection of Papers*. Retrieved from http://tncampuscompact.org/files/genref12.pdf.

Terenzini, P. T., Springer, L., Yaeger, P. M., Pascarella, E. T., & Nora, A. (1996). First-generation college students: Characteristics, experiences, and cognitive development. *Research in Higher Education, 37*(1), 1–22.

Torres, V. (2004). Familial influences on identity development of Latino first-year students. *Journal of College Student Development, 45*, 457–469.

Towner, T., & Muñoz, C. (2011). Facebook and education: A classroom connection? In C. Wankel (Ed.), *Educating educators with social media: Cutting edge technologies in higher education* (Vol. 1, pp. 33–57). Bingley, UK: Emerald.

Tym, C., McMillion, R., Barone, S., & Webster, J. (2004). First-generation college students: A literature review. Retrieved from http://www.tgslc.org/pdf/first_generation.pdf.

U.S. Department of Education, National Center for Education Statistics. (2013). *National Postsecondary Student Aid Study*. NPSAS: 2012. Powerstats.

Vaccaro, A., Adams, S. K., Kisler, T. S., & Newman, B. M. (2015). The use of social media for navigating the transitions into and through the first year of college. *Journal of the First-Year Experience & Students in Transition, 27*(2), 29–48.

Valenzuela, S., Park, N., & Kee, K. (2009). Is there social capital in a social network site?: Facebook use and college students' life satisfaction, trust, and participation. *Journal of Computer-Mediated Communication, 14*, 875–901.

van Oostveen, R., Muirhead, W., & Goodman, W. (2011). Tablet PCs and reconceptualizing learning with technology: A case study in higher education. *Interactive Technology and Smart Education, 8*, 78–93.

Vasquez-Salgado, Y, Greenfield, P. M., & Burgos-Cienfuegos, R. (2014). Exploring home-school value conflicts: Implications for academic achievement and well-being among Latino first-generation college students. *Journal of Adolescent Research*, 1–35. doi:10.1177/0743558414561297.

Violino, B. (2009). The buzz on campus: Social networking takes hold. *Community College Journal*, *79*(6), 28–30.

Vuong, M., Brown-Welty, S., & Tracz, S. (2010). The effects of self-efficacy on academic success of first-generation college sophomore students. *Journal of College Student Development*, *51*, 50–64. doi:10.1353/csd.0.0109.

Walpole, M. (2003). Socioeconomic status and college: How SES affects college experiences and outcomes. *The Review of Higher Education*, *27*(1), 45–73.

Warburton, E., Bugarin, R., & Nuñez, A. (2001). *Bridging the gap: Preparation and postsecondary success of first-generation students*. Washington, DC: National Center for Education Statistics.

Watts, D. J. (2004). The "new" science of networks. *Annual Review of Sociology*, *30*, 243–270.

Wellman, B., Quan–Haase, A., Boase, J., Chen, W., Hampton, K., Díaz, I., & Miyata, K. (2003). The social affordances of the Internet for networked individualism. *Journal of Computer–Mediated Communication*, *8*(3). doi:10.1111/j.1083–6101.2003.tb00216.x.

Williams, D. (2006). On and off the 'Net: Scales for social capital in an online era. *Journal of Computer–Mediated Communication*, *11*(2), 593–628.

Wohn, D.Y., Ellison, N. B., Khan, M., Fewins-Bliss, R., & Gray, R. (2013, April). The role of social media in shaping first-generation high school students' college aspirations: A social capital lens. *Computers & Education*, *63*, 424–436.

Yazedjian, A., Toews, M. L., Sevin, T., & Purswell, K. E. (2008). "It's a whole new world": A qualitative exploration of college students' definitions of and strategies for college success. *Journal of College Student Development*, *49*(2), 141–154. Retrieved from http://doi.org/10.1353/csd.2008.0009.

Yin, R. K. (2009). *Case study research: Design and methods* (4th ed.). Applied Social Research Methods Series, 5. Los Angeles, CA: Sage.

Yosso, T. (2005). Whose culture has capital? A critical race theory discussion of community cultural wealth. *Race Ethnicity and Education*, *8*(1), 69–91.

Index

About the Authors

HEATHER T. ROWAN-KENYON is associate professor of higher education in the Educational Administration and Higher Education Department in the Lynch School of Education at Boston College. She received her PhD in Education Policy and Leadership from the University of Maryland, an MA in College Student Personnel from Bowling Green State University, and a BS from the University of Scranton. Her research focuses on college access and student success, and student learning. Her work appears in publications including the *Journal of Higher Education*, *The Review of Higher Education*, and the *Journal of College Student Development*.

ANA M. MARTÍNEZ ALEMÁN is professor and Associate Dean for Faculty and Academic Affairs at Boston College in the Lynch School of Education. She is co-author of *Online Social Networking on Campus: Understanding What Matters in Student Culture* and co-editor of *Critical Approaches to the Study of Higher Education* and *Women in Higher Education: An Encyclopedia*. Her scholarship has appeared in the *Journal of Higher Education*, *Teachers College Record*, *Educational Theory*, *The Teacher Educator*, *Feminist Interpretations of John Dewey*, *Educational Researcher*, *Feminist Formations*, and *Review of Higher Education*. She is the editor of *Educational Policy*.

MANDY SAVITZ-ROMER is senior lecturer on education at the Harvard Graduate School of Education. Her scholarly interests include college and career readiness, school counseling and student supports, first-generation college students, social emotional

and K–16 partnerships. She writes and speaks widely
schools, universities, and community-based pro-
and deliver effective college-ready programming.
hor of *Ready, Willing, and Able: A Developmental*
ge Access and Success. She holds a PhD in Higher
on from Boston College.